UNDERWATER NATURALIST

Neville Coleman

ACKNOWLEDGMENTS

The majority of the photographs in this book are from the Australasian Marine Photographic Index a comprehensive scientifically curated visual identification system containing over 100,000 individual transparencies covering almost every aspect of aquatic natural history. With over 12,000 species recorded a large number are actually cross-referenced with specimens housed in Australian State museums and/or scientific institutions. This resource collection is the largest of its type in the Southern Hemisphere. However, regardless of my enthusiasm and dedication for the subject matter, without the assistance of curators, research associates and diving colleagues this book would have been a lesser achievement. Many thank yous to all. My thanks also to Anne Thomson for her typing skills and to Christine Glenwright of 'Artrix' for her computer artwork skills.

MALDIVES:	Mustag Hussain and friends; Jorg Aebi, instructor, **Dhigufinolhu Island Resort**
JAPAN:	Toshikazu and Junko Kozawa of **Anthis**; Mr Hajime Masuda and staff of **Izu Oceanic Park Dive Centre.**
PAPUA NEW GUINEA:	Dik Knight and staff, **Loloata Island Resort**; Max Benjamin and staff, **Walindi Plantation**; Alan Raabe, **MV Febrina** and staff; Rob van der Loos of **MV Chertan** and staff.
SOLOMON ISLANDS:	Jill and Grant Kelly, **Uepi Island Resort**.
INDONESIA:	Special thank you to Takamasa and Miki Tonozuka of **Dive and Dive's Co Ltd** of Bali for their expertise and friendship.
USA:	Paul Humann and Ned Deloach; NASA; Sandi Smith
QUEENSLAND:	Bob and Dinah Halstead, **Heron Island Resort; Heron Island Research Station; Lizard Island Research Station**; Peter Davie, **Queensland Museum**; Roger Steene; Steve and Dorothy Hills of **Pro Dive Milton**; Steve Parish; Dr Pat Mather; Dr Peter Saenger; **GBRMPA**, Dr Sandy Bruce; Dr Carden Wallace; **AIMS**
NEW SOUTH WALES:	Pat De Groot, **Pro Dive Travel**; Brian Busteed, **Howea Divers; Australian Museum**. **PADI Asia/Pacific**; Shane Ahyong; Dr Lesley Newman; AUF; Dr Alan Millar.
VICTORIA:	Bob and Margaret Burn; Rudie and Alison Kuiter, **Zoonetics**; Dr Jan Watson. **Museum Victoria**. Dr Tim O'Hara; Ken Bell; Dave Staples, Mark O'Loughlan; **CSIRO Publishing**, Phil Bock.
SOUTH AUSTRALIA:	Scoresby Shepherd; Ken Lang, Paddy O'Neill and staff, **Sea Suits (Aust). South Australia Museum.**
WESTERN AUSTRALIA:	Peter and Leslie Hudson, **Esperance Dive**; Ron Moore, **Albany Scuba Diving Academy**; Wally Rowlands; Loisette Marsh; Dr Diana Jones; Dr Gary Morgan; Dr Gerry Allen
NORTHERN TERRITORY:	Dr Phil Alderslade; Dr Chris Glasby.

Special thank you to Paul Humann and Ned Deloach, Reef Creature Identification (Florida Caribbean Bahamas) **New World Publications** for permission to use many of the line drawings from their books. This courtesy has added a greater degree of comprehension to this publication.

Line drawings and figures are acknowledged from many scientific and sources old and very new, and clearance has been obtained wherever required. Many thanks to everyone for your cooperation and assistance towards this educational endeavour.

Publisher: Neville Coleman's Underwater Geographic Pty Ltd
ACN 002 043 076

National Library of Australia
Cataloguing-in-publication data
Neville Coleman
Underwater Naturalist
New Edition
Full Contents/Index
ISBN 0-947325-29-8
First published July 2004
© Individual photographers as credited
© Neville Coleman, concept, design and all uncredited photographs
PO Box 702 Springwood
Qld 4127 Australia (07) 3341 8931
Fax: (07) 3341 8148
Web address: www.nevillecoleman.com.au
E-mail: worldofwater@nevillecoleman.com.au

Film by Kodak.
Scans by Kodak.
Nexus camera housing.
Nikon & Canon cameras.

2

CONTENTS

Dedication 4

Preface.. 5

INTRODUCTION

The aquatic environment 6

The new age of diving.....................................10

Environmental diving 12

Caretaking 14

Divers role in marine science 16

Eco-tourism...19

The future 21

Sorting out the stuff!......................................22

Simple classification 22

Understanding taxonomy....................................23

New species 24

The nature of things.......................................26

BASIC IDENTIFICATION

KINGDOM MONERA

Cyanobacteria:...27

KINGDOM PLANTAE

Algae:..28

Seagrasses:...30

Mangroves:...31

Phytoplankton:...32

KINGDOM PROTISTA

Forams:..34

KINGDOM ANIMALIA

Invertebrates

Sponges:...35

Cnidarians:
Hydroids, sea ferns, hydrocorals, fire corals,
siphonophores, sea jellies, box jellies, sea fans,
soft corals, sea whips, sea pens, sea anemones,
zoanthids, corallimorphs, stony corals, black
corals, tube anemones 37

Comb jellies:..53

Flatworms:...54

Ribbon worms:...55

Polychaete worms:
Leeches, tubeworms, bristle worms,
scale worms, spaghetti worms.........................56

Spoon worms:..58

Sea spiders:...58

Crustaceans:
Barnacles, mantis shrimps,
amphipods, isopods, shrimps,
prawns, rock lobsters, hermit crabs,
squat lobsters, crabs.....................................59

Molluscs:
Chitons, univalves, opisthobranchs,
nudibranchs, tusk shells, bivalves,
cephalopods 65

Brachiopods:
Lamp shells 69

Phoronids:
Horseshoe worms 69

Bryozoans:
'Lace corals', sea moss...................................70

Echinoderms:
Feather stars, sea stars, brittle stars,
sea urchins, sea cucumbers 71

Vertebrates

Ascidians:
Tunicates, sea squirts...................................76

Sharks and rays:..78

Bony fishes:...79

Marine reptiles:
Sea turtles, sea snakes, crocodile 80

Marine mammals:
Dolphins, whales..82

Project AWARE..84

View from the board;
Jean-Michel Cousteau......................................87

Project AWARE Board of Governors.........................87

Conservation..88

Physical land and water facts 90

Glossary 92

Index 94

World of Water bookshelf 95

Author's profile...96

"Part of the whole"

DEDICATION

PHOTOGRAPHY IS TO ME...

Pictures communicate, as an environmental photographer I see everything as important and endeavour to capture the meaning of life in a way everyone can share and understand.

Photography to me is more than pictures, more than film, more than cameras. Photography is life itself. It is this sharing of life that is more important than all the pictures in the world.

If I can open eyes, minds and hearts with words and images and let others share in the beauty of nature, perhaps they too will find their own reasons for being, and strive to bring about a greater understanding towards our seas, our shores, our selves.

Therefore, I really don't have any favourite pictures, or favourite animals, because I am seeking to portray parts of the whole, rather than singular entities.

The pictures themselves are not really mine, they are but visual expressions of life forces. The images are already created, all I do is interpret with a little technique and a lot of love.

Photography is a wonderful manifestation of self. A basic camera exposes a negative, yet produces a positive, a marvellous philosophic concept.

Words and pictures are the most powerful weapons we have to combat the environmental woes of the world. They put hope back into hearts, spirit the souls and light fires in the minds of apathy and ignorance.

Every day and night we are touched by the essence of nature; imagery, allows that touch, to linger...

AQUATIC IMAGERY

PREFACE

We are constantly surrounded by a multitude of things all taken in through our eyes and processed by our brains to form images through which we recognise and understand our environment, measure our existence and interpret all the aspects necessary for our lifestyles and ultimate survival. Every day and night our eyes and brains scan thousands of images which are recorded automatically; the intensity of each recorded image is determined by the amount of value we place on what is seen. In effect, it depends on the power of the specific visual stimulation by which we store or discard our visual memories.

However, no matter how fantastic our vision may be, or how brilliant the images and experiences, we see them only for ourselves. Because we grow up surrounded by different circumstances, things that are always there often appear trivial. *Sometimes it takes somebody else's vision to enable us to really see the nature of what is around us and recognise the beauty of what is within us.* Once the 'just looking' is replaced by the *art of seeing*, the entire perception of nature becomes an exciting new world and a lifetime adventure in learning begins.

We all have the ability to create. As children, we are naturally creative but more often than not the 'big people' misinterpret our sources of inspiration and early attempts are often judged as mischievous. Anybody who can remember their first venture at decorating the lounge room walls with crayons can vouch for this. Emphasised by red backsides and bashed ears, our survival instincts lead us to be more confined in our approach.

For others, the seed of talent is just waiting to be recognised and watered with encouragement. *Art* is amongst the highest forms of self-expression, and for as many 'selfs' in the world, there may be as many expressions.

Yet so few ever pursue beyond the edge of their momentary visions, or have the ability to caress a medium with their souls and turn captured perceptions into the magic of meaning. Few have the ability to translate, to infuse *nature* with *human nature* in such a way that *everyone* can share and understand.

Since the beginning of human evolution on this planet, beings have recorded forms and events of the world around them. As skills and intellect developed so too did the urge for communication. With each new rise of civilisation, culture grew and the art of making, telling, preserving, appeasing, recording and enshrining led to works on a scale both intricate and enormous. Yet, for all human preoccupation with art for art's sake, there is an art form so far beyond the fragility of frame that its very existence is within us all. Its presence shows we are but interpreters, 'creators' of creation, copiers of kind, for what we have accomplished is nothing when compared to *what already is!*

The many forms by which art is depicted and administered, along with each series of impressions and expressions, are more or less *interpretations* regarding the measurement of art. One can no more measure art than one can measure love.

To me there are three main forms of art.

The *Art* of Reality   –   *is*   –   The World of Nature

The *Art* of Expression   –   *is*   –   The World of Human Nature

The *Art* of Invention   –   *is*   –   The World of Imaginature (NC 1993)

The quote, *'Beauty is in the eye of the beholder'*, is quite possibly one of the greatest truths of all time, yet even more so, when applied to imagery in all its many forms. It has such an important part to play in rekindling the values and enhancing our understanding of nature.

INTRODUCTION

Not everybody can go to the moon, or observe the fascinating creatures in the hot thermals of the abyssal rifts (deep trenches in the ocean floor). Not everybody has the experience, or knowledge to mount expeditions to far off places and make discoveries on their own and bring back exciting new species, observations and photographs. Yet everyone can have an appreciation of nature, and for those seeking the spirit of the universe there is no greater adventure than exploring the **world of water** at our very doorsteps.

Perhaps we are drawn to the seashore for more reasons than we really understand, yet the understanding of 'getting back to nature' is within all of us. It is something which everybody can enjoy, from the young, to the young at heart.

Living throughout the Asia/Indo-Pacific area we are all in the same unique boat, islands surrounded by water where the largest oceans of the world pound in day after day, year after year. Yet even though 100,000,000 people may live within 50 kilometres of an aquatic or coastal environment, we know surprisingly little of what makes up this huge and vitally important area which covers over 80 per cent of our planet.

In some ways our ignorance is quite natural, for as land beings, we are more or less only familiar with what is around us in our day to day lives. Most people relate to the elements and seasons, the sun and the rain. Trees and most plants are upright and commonplace; their flowers have bright, attractive colours. Terrestrial animals of our lands are known to most. From a very early age we watch them on TV documentaries. Even the smaller backyard denizens such as snails, grasshoppers, cockroaches, flies, mosquitoes, butterflies, lizards and frogs are commonly encountered. We grow up with them; they are there.

For those seeking the spirit of the universe there is no greater adventure (Beqa Lagoon, Fiji)

Explorers in another world. Sharing the underwater experience enriches our lives.

With the **world of water** it is different. We go only as visitors so our time there is limited. We generally go for pleasure, or to relax, to enjoy ourselves, to participate in sport, or in the pursuit of food. There is usually little time left over to devote to looking at the sea and shore and relating to the plants and animals there as living things. Because of this lack of familiarity much of the true beauty of the **world of water** is hidden behind the blinds in our minds, our ignorance.

The majority of animals hide from predators and the heat of the day in nooks and crannies and beneath the sand and mud. It takes more than a little interest to find and observe them. With the exception of birds and the larger, marine animals such as whales, sea lions, dolphins and sharks many aquatic organisms are relatively small (nudibranchs) and although beautiful, may be well camouflaged.

Yet, it is the retiring habits and nocturnal behaviours of sea creatures that make them all the more interesting. Every aspect of nature's benevolence we can relate to and recognise, enriches our lives. **Gaining knowledge is like fertilising the mind, the more we know...the more we grow. (NC 1979)**

SCIENTIFIC RECORDING IN THE PAST

The first recordings of marine animals and plants and many of the early descriptions still in use were based on intertidal specimens, or those washed up on the shore. Some species were known only from their remains left after cyclones, or rough seas. Many sponges, algae, shells, fish, sea stars and sea fans were first recorded from tidelines. Storms, pounding surf, churning waves and the scouring effect of currents and high spring tides dislodge innumerable organisms from their habitats and cast them up on the tideline. Some creatures are often intact but others may be only a part of their original beings, and as the sun dries and shrinks them they no longer resemble their living shapes. Many are still only recognised by their dead or preserved remains.

Over the last 40 years, modern colour photography (film or digital) has developed as an important tool in recording and establishing the true forms, colours, patterns, designs, behaviour and relationships of living marine life.

Underwater imagery is the most efficient method of recording living marine life.

Marine scientists working in the field have an immense task (AIMS) (GBR)

As we explore the sea by beachcombing, reefwalking, snorkelling or scuba diving, we are expanding the boundaries of understanding and our perceptions as beings on this planet. However, it must be remembered at all times that the ocean is one of the most relentless and powerful forces on earth. There are no means of beating it, we can only survive on it, around it, or in it by learning its many moods and patterns and making decisions based on that knowledge.

It is going to take many years to record and understand even the minimum amount of information necessary to achieve a working knowledge of the oceans and their inhabitants. Marine science institutions need help, they need underwater people with a sense of adventure, and a pride in personal achievement and commitment to assist in the gigantic task ahead.

Pioneering the **world of water** isn't easy; it takes dedication, patience, perseverance, understanding, sacrifice, money, hard work and an awful lot of love. The rewards may not be monetary rewards, but rewards in the truer sense, a richer fulfilment of knowledge and understanding, the rewards of you giving your 'go' a go and discovering the adventure of life by sharing in the nature of things.

This course will introduce you into a world of **adventure**, a living, breathing life story, an educational experience, yet, by its existence, a tribute to every naturalist who ever lived. All the so-called crazy eccentrics whose writings inspired my childhood imagination and sent me off on a journey of exploration. From collecting oysters, mussels and periwinkles for food in the Lane Cove River in Sydney, New South Wales, to photographing over 12,000 species across the far reaches of the **world of water** and discovering over 500 new species of sea creatures. From sea shore to sea floor, the ultimate discovery...welcome to the journey.

The World of Water is everywhere and the pursuit of knowledge infinite (Qld) (N.C.)

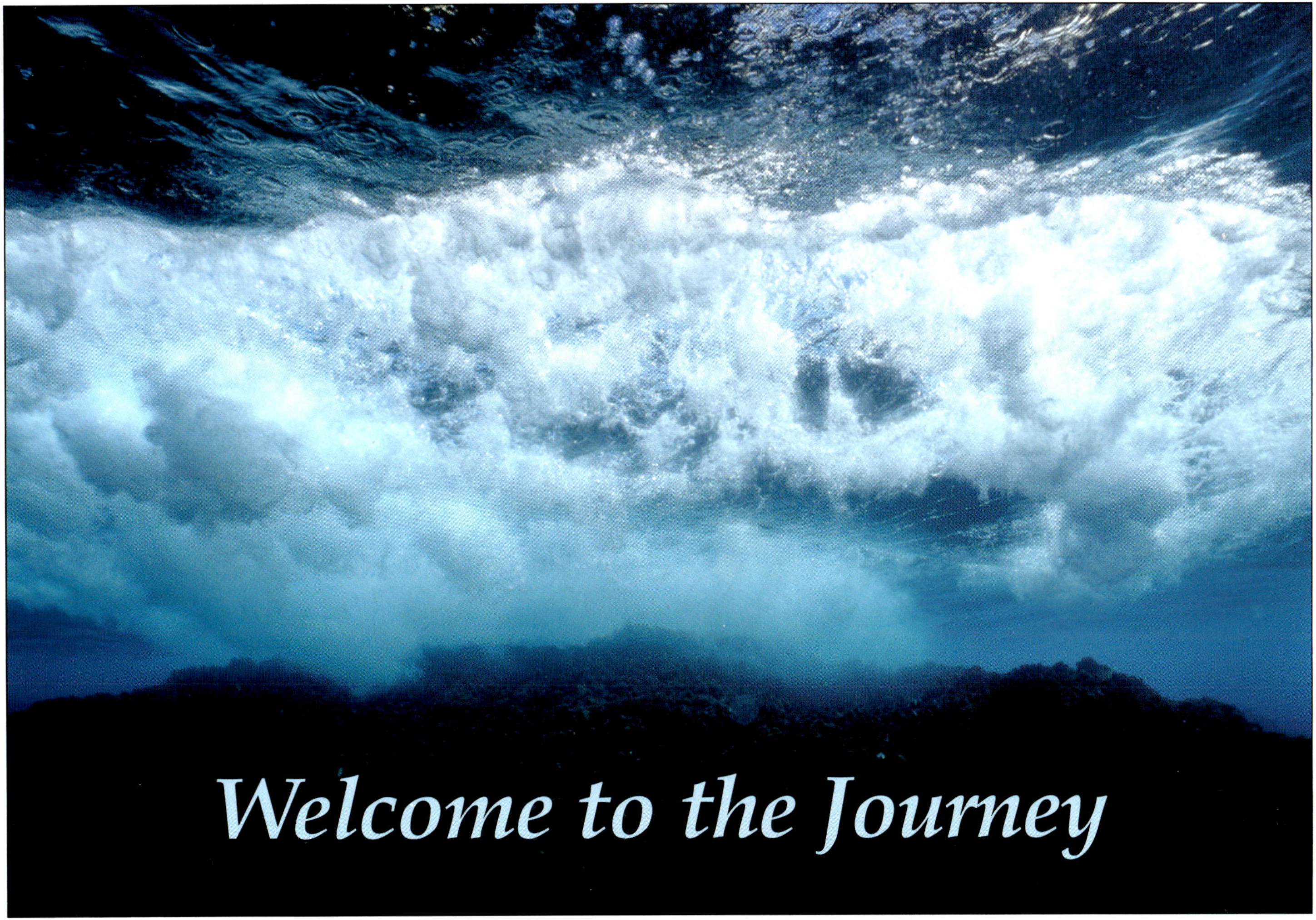
Welcome to the Journey

THE NEW AGE OF DIVING

What diver hasn't at some time been suggestively influenced as to the dangers of man-eating sharks, razor-toothed barracuda, ferocious moray eels and giant cod? These negative fears are often forced upon us by movie producers, writers, friends, parents, spouses, other divers, or whoever happened to be in our scene at the time. The promotion of these fears are for the most part, pure fantasy.

Sharks are an important part of the aquatic environment, most are non-threatening to divers

Our dive courses extend our knowledge and confidence

Of the 335,000 life forms inhabiting the oceans of the world, from tiny microscopic dinoflagellates to the mighty blue whales, there are only a small percentage which adversely affect humans. (Read *Dangerous Sea Creatures* by the author or come along on a **Sea Creature Safety... Dangerous - Venomous - Poisonous (Prevention is better than cure specialty course).**

Humans are not the normal, natural prey of marine creatures, yet in the past we have considered ourselves (due to extraordinarily irresponsible media exposure) to be on the menu of every creature in nature. **In short, humans have always been at war with nature.** Our ignorance and fear have bestowed totally ridiculous powers to many of these creatures, when in truth most have no understanding whatsoever of our existence (most cannot even see us!).

Those animals which inhabit the **world of water** (once an unknown and fearsome place) are only beings, in many ways similar to ourselves. By knowing, recognising and understanding them a little more in their world, we can go amongst them with respect, rather than fear, wondering at such fascinating creatures with their amazing abilities to defend themselves against a host of predators. That these astounding devices sometimes work on inexperienced human visitors is unfortunate.

However, a diver who is not afraid is a fool. In diving, 'fear' is an important necessity. I am talking about the subtle fears of reality, the continued awareness and realisation that beneath the water you are in a world other than your own. To enter this world and survive you must abide by certain procedures.

Scuba diving training courses have strict rules for survival which if applied correctly enable the certified diver the opportunity to explore and enjoy the **world of water**. The more advanced a diver becomes in experience and knowledge the greater the enjoyment this adventure activity provides, *but* at all times a **diver must be aware!**

Experience

Experience is all a matter of interpretation, it just depends on which side of nature you are interested in. If you measure yourself against divers with twelve or twenty years behind them, your twelve hours, days, weeks, or months may appear a little insignificant.

This is not so. The number of years, the dive locations, or the 'zillion' hours bottom time, prove nothing. When it comes to recognising and understanding marine animals in their natural habitat, inexperience is something **all** divers have in common.

So, regardless of how minimal your bottom time is, you have the opportunity of learning more about your adopted environment during this course in a few hours than many of your counterparts have in 5, 10 or 20 years of swimming around with their eyes shut.

No doubt you have heard the expression **'the sea is the last frontier to be conquered'**. In one aspect this is indeed true, the sea is the last frontier, **but why does it have to be conquered? Perhaps, because in the past, 'killing was easier than understanding'?**

The Challenge

Throughout the Asia/Indo-Pacific area, diving scientists are ever-increasing and the old days of complete dependence on amateurs is on the way out. You may ask, 'what then is left for the interested amateur?' Nothing much I'm afraid, just the life histories of 30 or 40,000 animals, hundreds of new species to be found, and the thousands of ranges to be recorded and extended. We are only at the threshold of learning in regard to marine natural history; it will take many thousands of people several centuries before we can hope to evaluate the sea and its secrets. It's all out there, waiting to be discovered.

To discover, one must first know where and what to look for. As you proceed on this course some of these mysteries will be revealed.

Understandably it will not teach you all you want to know, but for those of you who are keen it will give a basis on which ideas can be formed to further your own particular interests.

By the application of knowledge and the extensive use of underwater and applied photography utilised throughout this course, the student will become visually acquainted with the sea, its creatures and some of their habits and behaviour.

In many ways the reefwalker, snorkeller or diver will become aware of him or herself as an explorer in another world. Eyes that are used to record the familiar everyday things of terrestrial existence must be retrained to recognise the unfamiliar and unusual. The unbelievable array of colour, shapes and patterns which at a quick glance may mean nothing, must be singled out, established in the mind and remembered. To get the maximum enjoyment out of this adventure activity one must become a 'disciple of discovery' and see with eyes that are open.

A maze of coral heads in Wistari Reef lagoon (GBR)

A coral reef in its entirety is an amazing structure as a single entity, but the ability to recognise, relate to and understand the thousands of individual creatures which make the reef what it is, increases its wonder a thousand-fold. Once the light of knowledge has the opportunity to glow, the **world of water** and all its creatures become a **world of wonder** and a source of amazement for ever.

Orientation - Environmental Diving

Because we are normally land-orientated beings, the familiar everyday things around us are observed and recorded almost unconsciously. Land plants and animals are mostly accessible and those which are not we can almost always find reference to in the multitudes of books available.

With the sea, it is different; the majority of organisms are unfamiliar to us. Their shapes, colours, habits and environment are completely new and therefore a certain amount of training is necessary before we can begin to learn their secrets. The first and most important requirement is **to learn to exist comfortably** in this new environment. To be able to spend the maximum amount of time on the bottom, free of any hang-ups, real or imaginary.

Only when a diver is completely relaxed and moves over the sea bed slowly and methodically, will he or she be able to recognise its inhabitants. A lot of new divers see very little because they are swimming around endeavouring to cover as much territory as possible or using up all that precious air and reducing the visibility to zero by disturbing the bottom sediments with sweeps of their arms and clumsy fins.

Unsecured consoles can be dangerous to the diver and damaging to fragile corals (GBR)

Even with a heavy camera housing this photographer shows perfect buoyancy control (GBR)

Diver control

One of the most important skills necessary in becoming an underwater naturalist, and even more so, when aspiring to become an underwater photographer, **is diver control.**

It is plain common sense when we are experiencing the wilderness in a terrestrial national park that we don't chop down the trees, four-wheel-drive off-road, dig up the flowers, set alight the countryside, pollute the waterways, shoot the fauna (to get a closer look), feed the animals poison food, or otherwise mistreat the environment the thousand and one ways humans can.

In the **world of water**, many life forms are much more fragile than we are used to on land. Some animals, such as flatworms, some sponges, nudibranchs, ascidians, sea fans, soft corals, sea anemones, feather stars, brittle stars, sea cucumbers, comb jellies, bryozoans and sea spiders, are so fragile that just the act of picking them up can seriously damage them.

Defence Mechanisms

Animals such as sea urchins, blue-ringed octopus, sponges, fire corals, hydroids, crown-of-thorns sea stars, sea jellies, cone shells, porcupinefish, toadfish, scorpionfish etc, may in turn seriously damage a clumsy diver.

Most divers have little idea of their presence or impact when underwater, and most are unaware as to what they represent to the creatures. When looked at realistically, an averaged-sized diver when horizontal with arms outstretched, being propelled by 1 metre long fins (40 in), may be very close to 3.50 metres in length (11ft) which means that with the exception of cetaceans, sharks, marlin, dugong and manta rays, divers represent huge physical proportions to the majority of sea creatures.

Three metres of wet-suited diver can represent a fearsome presence to marine creatures

Of course, the fact that the greater numbers of invertebrates do not have focussing eyes does not mean that some do not hear us, smell us, feel us, or otherwise sense our presence. Others, of course, appear completely oblivious to our being unless we touch them, kick them, stand on them, break them, collect them or bump into them. Even then they have no idea what we actually are - it's just a physical contact that might come from any other sea creature or environmental factor such as storm damage or predators.

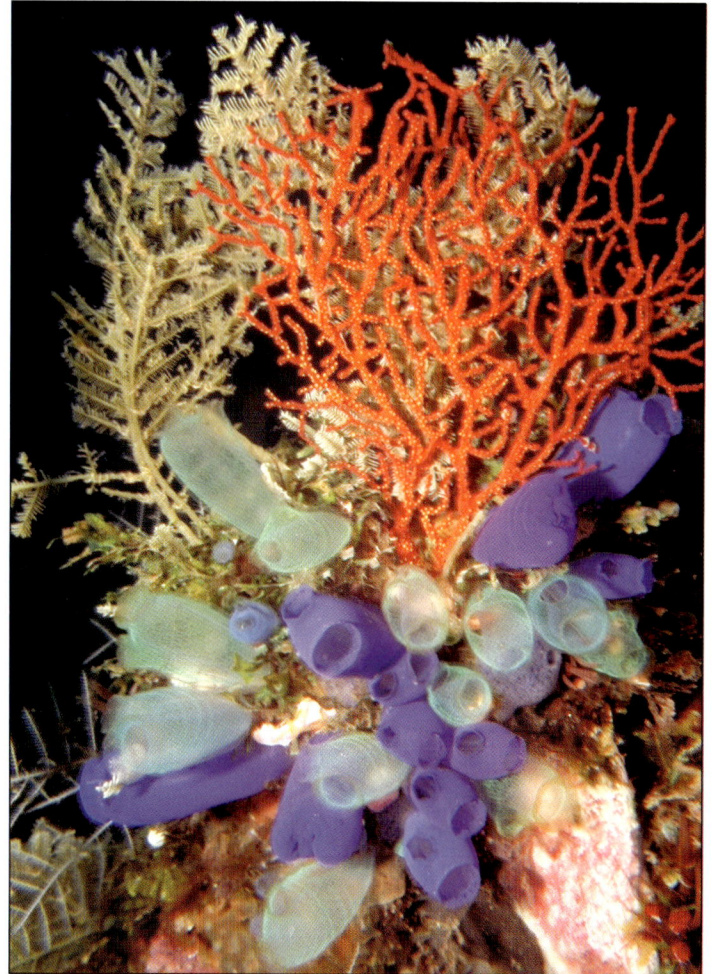

An underwater torch brings out the true colours and beauty of sea creatures (Indo.)

However, it can be pointed out at this time that in nature, on land, in the air, or in the sea, very few animals are clumsy. Not many birds misjudge and bump into trees in full flight, even when speeding through dense rainforests, or thickets; they are incredibly agile and orientated. The same may be said about our terrestrial wildlife even when scared and in full retreat away from gun-wielding humans, they rarely misjudge footing or come in contact with rocks or trees.

Under normal conditions, I've never seen a clumsy fish or a clumsy whale or a clumsy dolphin, though I must admit hermit crabs are not too orientated sometimes. But to them it doesn't matter if they drop off a rockface, they have their shell house to protect them when they fall.

One could argue that elephants and rhinoceros aren't very aware when they crash through forests or destroy groves of trees, and this is true. It appears that the only animals that behave in this way are protected by thick hides. Their nature can be destructive to their environment but they themselves are not unduly affected.

In turn, visualise a clumsy diver in full wetsuit with gloves and solid plastic fins...also protected. Does this not conjure up a similar scenario?

Yet the **world of water** is so huge how could a few clumsy divers make a difference? In the true understanding of the universe and the pure aspect of physical breakages, little difference at all. There is more physical damage done to the reef during one storm than the entire population of continents could cause jumping up and down on the exposed reef at low tide. For in the next coral spawning, new corals will settle and begin the cycle all over, quite oblivious as to whether it resulted from storms, or people. Most corals are colonial animals, and as such when damaged, or even broken, they do not die. If the broken piece falls into a suitable position (not in the mud), it will continue to grow into its own colony structure, as shown by recent successful experiments into transplanting corals on the Great Barrier Reef.

Caretaking

So why is it so important to ensure that diver control is maximised and physical damage is minimised?

1. With more and more humans being introduced to the **world of water** and hundreds of thousands of dives are taking place per day, especially in many marine park areas. For example, on the Great Barrier Reef, dozens of giant catamarans can each transport 300 to 400 people to the outer barrier pontoons in a single day. Dive sites such as the Cod Hole are dived regularly by hundreds of divers from any number of dive boats. Unless **divers are made aware of the damage they can do** and the instructors guide their actions, each one of these regularly visited areas could become a 'bomb site' as uncontrolled, unaware 'divers for a day', accidentally can reduce the fantastic living architecture of the surrounding coral to rubble and spoil the experience for other divers.

The outer Great Barrier Reef pontoons; personnel on the giant catamarans do a wonderful job providing real underwater experiences to thousands of tourists. The reefs are well managed and show little (if any) damage

Hopefully we are learning from the Americas, and taking a conservative approach towards our eco-tourism diving. By creating a **common sense awareness** and an educational **understanding of the marine environment** in the mind of each new diver, we can avoid the mistakes of the past. Maintain a protective role in the present, and guard against the increased activities of the future.

2. By expanding on and implementing the concept of environmental friendly divers by practicing the skills and suggestions under the next heading, everybody is better off - the ocean and the diver. With better body orientation, **better buoyancy control**, better understanding of the habitats and the marine creatures themselves, better storage of consoles, better breathing control, better movement management, better stabilisation, better, more personalised and comfortable equipment selection, better attitudes and realistic dive plans, **better diving procedures** turns out **better divers**.

Even more important, once these skills and understandings become common practice, it is amazing how much more a diver will see and experience on every dive. Without pounding around as thick-skinned 3.50m (11 ft) berserk mini-subs intimidating even the bravest of marine creatures by chasing, grabbing, poking, bubbling, standing, sitting, lying, kneeling, kicking and crunching around, over, amongst and through the seascape, it's amazing what comes out to view the visitor.

It is even more amazing what goes on around the diver who moves slowly along above the substrate, using minimum propulsion, expelling minimum exhaust bubbles. Or just hanging in liquid space vertically, or horizontally, arms together observing, one can experience the creatures going about their every day lives, as if the visitor doesn't exist.

By **observing**, and **recognising** what goes on around them, divers can enhance their dive experience a thousand-fold, and by **employing a more controlled behaviour,** they increase the quality and longevity of popular dive sites and their inhabitants.

So, by becoming an environmental diver, minimum input and maximum control produces the best results for all.

Oblivious to the chaos

Video photographer shows a good understanding of balance and control (GBR)

Early exploration of the sea began with early exploitation. Scuba spearing decimated many areas

Diving scientists are increasing across the Asia/Indo-Pacific region

Today's divers have the opportunity to help advance World of Water knowledge (GBR)

Exploration

Divers role in marine science

For thousands of years the Asia/Indo-Pacific sea people explored, settled, fished and traded accross the oceans.

Over two centuries ago Australia and the South Pacific were rediscovered by seafaring Europeans who faced the unknown of their time and accepted the challenge of the sea. The frontier of our time is the same sea and its challenge has no more diminished with the passage of time, for it is beneath the sea that humans face a millennia of nescience.

Diving is a lust for life, and divers are on the cutting edge of new-age exploration. Nature is all around us, but nowhere is it so fantastically portrayed, or so unknown as beneath the sea (and nowhere is it portrayed more beautifully than throughout the Asia/Indo-Pacific region. With an upsurge of interest in environmentalism, a new wave of diving exploration is emerging throughout the world. Underwater naturalists, eco-divers, (call them what you will), are forging ahead, creating a whole new approach to the adventure activities of snorkelling, snorkel diving and scuba diving.

Make no mistakes with history; early exploration of the sea began with early **exploitation** of the sea, and it hasn't stopped since. Yet some of us have realised that unless we can alter attitudes and turn the 'taking', into '**caretaking**', the precious resource of species will vanish from our oceans, and once gone, nothing will bring them back!

Through the evolution of change, the old 'he-man' image so diligently portrayed by the manhood-proving blood sports we were brought up on, is declining. Competition is a fine testing ground, but it should not be at the expense of the environment, or its over-exploited inhabitants.

By altering our attitudes we can turn the taking into caretaking.

The **world of water** is an equal opportunity for all; sustenance fishing, amateur and professional fishing, divers, snorkellers, underwater photographers, naturalists, tourists, scientists and students, big kids and little kids; each country and its traditional landowners and Governments are in charge of the ocean's resources and these precious resources must be managed realistically.

It's very easy to become blasé and say 'it's a big country, let somebody else take care of it' and remain with a selfish attitude. Big or not, the figures show that we are winning, and the environment is losing. We must become **managers of ourselves** before we can become managers of the natural environment. Each person needs to be a protector. We need to be knowledgeable conservationists and understand our dependence on the nature of things. We must become **aware**, beginning with ourselves.

Conservation and effective management can not be successful through government bodies, or big-brother type political protection. Conservation can only be successful through **people, ordinary everyday people,** who bother to **care!**

Certainly in some areas there is improvement; at least the problems are being recognised and in some cases being acted upon. Attitudes are slowly altering *BUT!* there is *still a long way to go...*

Although gloves often make handling small fragile critters damaging, this diver has fanned the flatworm off the bottom and cradled it to share with her dive buddies (LHI)

On your first snorkel or dive, no doubt, you will see a number of things with which you are unfamiliar. It isn't just a matter of touching, or picking up everything in sight and examining it willy-nilly. Animals live in many habitats and often require careful handling even to be observed, or studied. It's quite useless picking up a small soft flatworm or nudibranch with gloves, or holding it tightly, as even the most careful approach will often squash the animal, or otherwise damage it.

Many-armed feather stars are very fragile and unless care is taken you may end up with a lesser-armed feather star. The most efficient way of visualising for study is to have a series of self sealing bags in your mesh bag. Then each animal, or series of similar animals can be carried separately while the water in the plastic bag cushions them, and after close examination, observation, or photography, they can be returned back to sea unharmed.

Eco-tourism has a marvellous opportunity to contribute at every level of participation (GBR)

Quite a lot of marine animals are nocturnal and during the day they hide in caves, under ledges, or beneath rocks. It is up to the naturalist to investigate all the nooks and crannies in order to discover some of the more elusive species. Because some species may be well camouflaged, many may remain unseen by you. There are three main ways in which animals and objects are noticed underwater -

MOVEMENT - COLOUR - SHAPE

Movement is the main way our attention is drawn to notice marine life

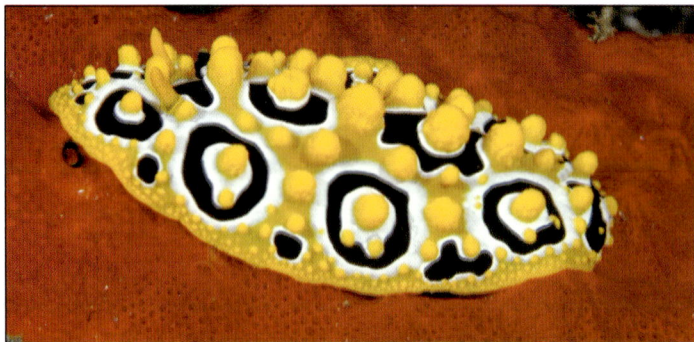

Colour is the next primary visual indicator; using a torch improves one's ability to see

(NSW)   Shape is used to determine visual identification of different groups   (PNG)

During your investigation, thought must be given to the environment and to the habitat of the animals. If rocks are turned, then they must be replaced to their original positions as gently as possible so as not to squash what is beneath them. If not, all the smaller invertebrates will be eaten by fish and those sessile dwellers of the dark places will be exposed to light and perish. In all, the underside of a rock is a micro-habitat and if left overturned, life on both sides will be forfeited.

It only needs a few careless people and whole areas can be affected, even when the rocks are underwater. This is also extremely important for intertidal investigations - **at all disturbed areas rocks must be replaced.**

Photographs of animals and plants mean little unless you know something about them. The only way to do this is to make your own observations while underwater. For this a good memory, a slate and/or a camera are necessary. It is also essential to keep a log book, or diary of each dive recording everything observed, photographed, or encountered. **PADI's adventure log book is a good example.**

Date	Dive Site	Surface Conditions
Depth	Visibility	Bottom Conditions (Habitat)
Observations	Dive Duration	Species Listing

This information can then be referred to at future times and quite often a pattern of behaviour can be established on a species. Only by **experience** and **investigation** can the correct interpretation of a creature's habits, or life history be recorded. To find a mollusc sitting on eggs doesn't always mean it laid them. Quite often some molluscs feed on the eggs of others. A sea star found on the reef may be clinging very tightly, but unless it has its stomach exuded, a feeding observation is not verified. It all takes time and effort, but for those who persevere, the rewards are endless.

Eco-tourism

Eco-tourism - brave new words for brave new worlds - has injected a breath of fresh air into the very soul of underwater education and although still in its infancy, it offers the opportunity for like-minded people to not only share the adventure of diving activities, but to take part in the discovery and documentation of our underwater wildlife's very existence.

There is little known and so much to learn and record in the field of marine natural history that **everything must be judged as important** and worthy of recording and reporting.

To me, a reefwalker, snorkeller or diver - be she or he a naturalist, photographer, scientist, observer, or adventurer - is an explorer. Not just an explorer in another world but an explorer in another concept, in which ideas, techniques, philosophies, languages and ideals are constantly being changed, remodelled and invented.

The world famous Potato cods at the Cod Hole off Cairns, Qld have been visited by many thousands of divers for over 20 years (GBR)

*Our planet Earth is in reality a **world of water** because all life depends on fluids*

The sea is only an extension of the land but it makes up 71% of the total surface area of our planet. Yet we call it, Earth. We cannot in any way, shape or form ever begin to understand this vast unknown part of our everyday existence by standing at a distance and throwing rocks, or by sticking pins in it to see if it's alive. Nor can we hope to evaluate the sea and its contents by remote control robots. We must go down and see for ourselves. **We must study and observe the creatures and their relationships within the total, overall balance of life as we know it.** It must be realised that the land, the air and the sea are all one gigantic system, all related, all connected, all interdependent and all important.

However, the **world of water** isn't mine, nor do my concepts belong to me. The issues I stand for, represent, believe in and fight for are everybody's issues, and everybody's responsibility. Unless we, as people, all make a stand for the aquatic environment and support the ideals of those whose dreams and commitments go beyond the ordinary, then what we **don't see** *is exactly what will be left.*

Knowledge is like a flight of steps through the evolution of time. We cannot begin to climb unless someone, somewhere builds the first step. Others come along at later stages using the first step, add their own knowledge and build another step. And so it goes on.

Just because somebody who builds the 50th step might know more (because they had 49 steps of somebody else's knowledge already built), it can never detract from the one who built the first, second or third steps, etc. We can always improve, even if it means going back to the step before, and starting again. Each observer of nature, be they in their backyards, or beneath the sea, if they are going to build a step with a good foundation, they must never be too stubborn, too proud, too selfish, bigoted, or too cowardly, to face the reality of truth. Being wrong is all a part of the learning experience, the understanding experience, the building experience. Wrong, is only wrong if we don't have the guts to admit it and learn by it. Make it right, then get on with life. We owe our **world of water** that much at least!

In 35 years of contributing to the steps of knowledge perhaps the job I set myself and the roles of explorer, educator and conserver were somewhat idealistic for the measure of one? Perhaps they were just the aspirations of a passionate dreamer who chose another way to fight for his beliefs, who created a reason for his being, a most important reason.

'The putting of nature *back*, into human nature...'

With your help, perhaps we will!

> *Humans are at best...their worst,*
> *and at worst...their best!*
> *Life is the equality of opposites*
> *Evolution by adversity*   N.C. 2004

The Future

The future of the **world of water** depends entirely on the world of humans. Unless we accept the responsibility and endeavour to change our attitudes towards its welfare **today**, the children of tomorrow will only mirror our ignorance. They deserve a better inheritance - after all, they have to live with, and attempt to correct our mistakes. They can only do this if we give them a legacy worthy of their belief. We must make peace with nature and heal the wounds of centuries. We must accept the responsibilities of our ignorance and **endeavour to change our eternally bad habits. We must alter our attitudes** and understand that this war with nature cannot be allowed to go any further.

Human nature, such as it is, can do anything it wants, but it must first be made aware of the circumstances the **world of water** is in; and it is not a pretty picture. According to all the right sources there are around 30,000,000 life forms on earth and we are losing 140 a day at our present rate. Yet with so many species unknown and still to be described and so many millions of species whose life styles, functions, natural histories are still a mystery, what we really need is an upsurge of interest towards nature and its importance to survival and well-being.

I realise that life is a struggle and that times are tough and I fear they are never going to get better. Products of our environment we may be, but we don't have a go through life packaged in the wrappings of the past, or a soul-destroying programming of the present.

Some could argue, "Learning about nature isn't going to improve my situation in life..." **yet nothing of substance is ever accomplished without enthusiasm and belief.**

If we realise it, nothing creates nothing; life is more than nothing. **"Each life force has the power of its being, but the power of being is only activated by the power of believing**" (N.C. 1980).

I believe that by taking good care of nature we are in essence taking good care of ourselves. To take good care of nature, we must understand nature. We need more people dedicated to that understanding.

Lady Elliot Island Resort (GBR) Underwater Naturalist Specialty Course students and staff displayed how educational eco-tourism could be applied

We need more observers.

We need more recorders.

We need more writers.

We need more photographers.

We need a new generation of naturalists to explore, observe, study, compile and publish what's around us.

We need to bring about a new **awareness** and fill our lives with the magic excitement of knowing. The only real treasure is living treasure, and we are custodians of the richest treasure trove in the universe - life! Surely, a little faith in it is not too much of a burden for us to carry, or protect. Whatever the afflictions of the human race, nature has the answer, it's just a question of **balance**.

"Time's a treasure, too soon it's spent

Life's not given...it's only lent..."(N.C. 1991)

SORTING OUT THE STUFF!

Simple Classification

There are thought to be between 10,000,000 and 30,000,000 life forms on Earth. Of those approximately 1,500,000 species have been described and named. However, as this course is primarily concerned with marine animals and plants, there are only a mere 335,000 estimated species to sort out.

One single rock garden at 30 metres off Lord Howe Island produced over 100 species on 3 dives

UNDERSTANDING SCIENTIFIC CLASSIFICATION

The basic classification method used in this course is a scientific one based on a hierarchal system, whereas the principles are arranged by descent from the highest (largest principle - **KINGDOM**) to the lowest (smallest principle - **SPECIES**).

THE KINGDOMS

For the purpose of this course four **Kingdoms** are included:

(1) **Kingdom Monera** - bacteria

(2) **Kingdom Plantae** - algae to seagrass

(3) **Kingdom Protista** - Forams

(4) **Kingdom Animalia** - sponges through to marine animals

ACCEPTED CLASSIFICATION SAMPLE

LEVEL	SCIENTIFIC TERM	COMMON TERM
Kingdom	Animalia	Animals
Phylum	Echinodermata	Echinoderms
Class	Asteroidea	Sea stars
Order	Spinulosida	Spinulosids*
Family	Asterinidae	Asterinids*
Genus	*Pateriella*	Pateriellids*
Species	*nuda*	Naked sea star*

* Note: Anglicised Latin terms may be used as common terms where no other word is in usage. Although this structure will apply to the arrangement of species within each phylum the book begins in the traditional scientific way by presenting the lower life forms at the present status of scientific knowledge.

ARRANGEMENT OF TAXONOMIC DIVISIONS

The arrangement in most cases refers only to the **phylum**, the **class**, the **family**, **genus** and **species**. Explanations have been kept as simple as possible.

Explanations

Phyla: A primary taxonomic division of animals and plants

Class: A comprehensive group of animals or plants ranking below a phylum

Order: A comprehensive group of animals or plants ranking below a class

Family: A comprehensive group of animals or plants related to each other ranking below order

Common name: Common names are those which are in general use, have already been published, or are directly based on the original scientific name interpretations. Where no such common name is available (for example, for undescribed species), terminology has sometimes been constructed from outstanding features of the species' external anatomy, colour, pattern, design or failing this, the locality where it was discovered.

Scientific names: It's important to distinguish one thing from another, to recognise, identify, talk about, teach or learn, animals and plants must have a reference point - a name. Common names, while providing a beginning are of little use in multiple languages or if several species on a worldwide basis have the same common name. In order to overcome this confusion in the middle of the 18th century the Swedish botanist, Carolus Linnaeus, published *Systema Naturae* and provided the basis of scientific nomenclature which is accepted throughout the world today. The concept was exceptionally well thought out. The names given to organisms when they are described and published scientifically are written in Latin. In this way it does not matter which language a book is written in, the organism's scientific name will be the same and recognisable throughout the world.

Each scientific name consists of two Latin words and is known as **'binominal nomenclature'.** the First name always begins with a capital letter (for example, *Cassis*) and is known as the **genera (which might be considered the organism's surname)**. The second name, which always begins with a lower case letter, **(and which might be thought of as the organism's given name)** is the **species** name (for example, *pelagicus*).

Generic names are unique within the animal kingdom but species names can be re-used to describe the animal in question. Thus the edible sea-urchin in the United Kingdom is *Echinus esculentus* and in the Caribbean is *Tripneustes esculentus*. It is customary to write the name of the author after the scientific name, for example, *Echinus esculentus* Linnaeus, and to add the date that this name was first published. If the species is transferred to another genus, the author's name is placed in brackets to show it has been re-assigned, eg, *Echinus esculentus* (Linnaeus, 1785).

Linnaeus' system was good because it indicated natural relationships. For example, the great cats *Pantheria leo* (lion) and *Pantheria pardus* (leopard) are obviously closely related yet they are not the same in appearance and distribution. The cheetah is not very closely related to either and is quite different in appearance and behaviour. It is placed in another **genus** *Acinonyx* with **species** name *jubatus*.

Linnaeus also grouped similar genera together into **families** and similar **families** into **orders**. There are several **families** of two-winged flies; the familiar house flies, the gnats and mosquitos, and the non-biting midges. Each of these three **families** is distinct and recognisable, but they all fall into the **order** Diptera - flies with two wings.

Likewise **orders** are grouped into **classes** and **classes** into **phyla**. Thus we end up with an hierarchical system which tells us about the supposed relationships and links between the animals we are dealing with. At the top is the **phylum**, containing all the animals supposed to have a common evolutionary origin. At the bottom is the **species**, a unique example of that **phylum**.

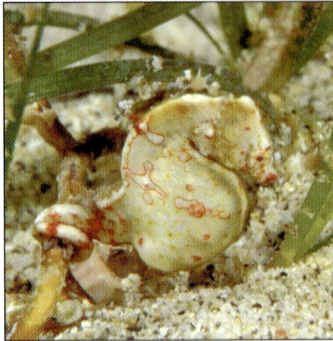

Coleman's pygmy seahorse Hippocampus colemani Kuiter, 2003 was discovered at Lord Howe Island in 3 metres

Holotype of Coleman's pygmy seahorse. Specimens housed at the Australian Museum (Photo: Rudie Kuiter)

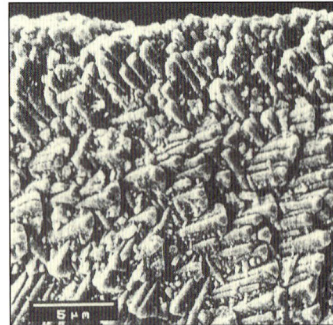

Coleman's soft coral Moolabalia nevillecolemani Alderslade, 2001 from Mooloolaba, Qld in 25 metres. Specimen housed at Northern Territory museum (microscopic spicules)

New Species

If an animal new to science has not had its name published, or if we are examining an animal for which we know the **genus** but not the **species**, it is referred by its accepted generic name followed by sp. which is short for **species**, for example, *Cassis sp.*

If we want to refer to a group of **species** within a **genus** we can use spp, for example, *Cassis* spp. (*Cassis* is a **genus** of large marine predatory snails known as helmet shells).

All the names in the hierarchy above the **species** name are there for the convenience of seeing how the **species** relate to each other. **Family names are used in some books as well as genera and species, because they help to provide a better understanding between the relationships of closely allied genera.**

Historic Confusion

In the 17th, 18th and early 19th century many lands were still being discovered. Expeditions from various countries visited out of the way places, collected specimens and took them back to their universities and museums. Because communications were so poor and literature almost non-existent, scientists from several lands often discovered the same animal. Hence, many of the same **species** received several different names, often in several different languages.

When an animal is named, the specimen described is called the **type specimen** and is lodged in a special section in museums. **Type specimens** are the most valuable of all specimens to an institute because of their reference value in identifying similar organisms.

Australia has a rather large problem, as many of its **type specimens** are lodged in museums in the countries whose expeditions discovered them. So to get an accurate identification on one of our animals, one must first track down in the literature when it was described, by whom, and where the **type** is lodged. It could be in France, Spain, England, or Germany.

To go through the records of perhaps 150 to 200 years to find out which country, museum, department, room, shelf and bottle that organism is in, takes a lot of time, effort and cooperation.

To compensate for these problems, scientists publish **keys** in their descriptions of certain groups of animals. A worker or interested party once knowing how to read and use the **key** can them eliminate certain factors and come up with an identification. Many aquatic organisms are **keyed externally**. That is, a series of recognisable external features are given to define a species. Examples for this are fish, in which shape, spines and scale counts are important, and shells and crustaceans, where the description of the external skeleton has been the chief recognising factor. However, other groups such as sponges and soft corals are **keyed** by **internal factors** (spicules).

This gives some insight into the real fact of why species identification in the past has in most cases, not been as simple as it might have appeared. The larger organisms the diver sees are all fairly well distinguishable to at least separating them into their particular **phylum**.

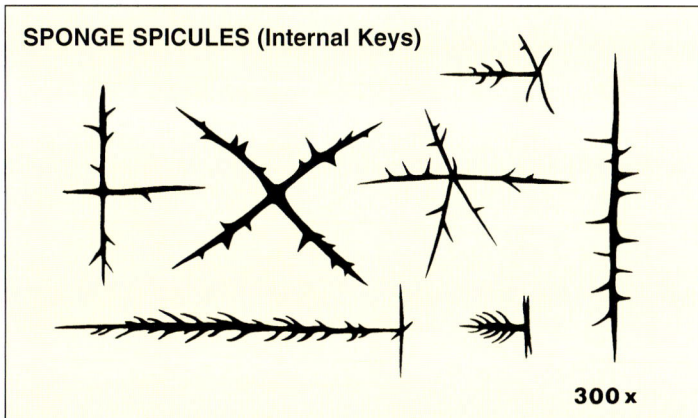

SPONGE SPICULES (Internal Keys)

300 x

To the new diver this small area of reef at Beqa Lagoon, Fiji appears daunting to comprehend. However, once explained, marine life becomes more than just "pretty stuff!"

It takes surprisingly little time for the student to grasp the basics of recognition once the initial features have been explained, good references and visuals supplied. The 'stuff' then takes on a whole new meaning and becomes a multitude of living, breathing organisms, each with an incredible story to be discovered and shared. As each organisms is recognised and understood, it takes on a new value and for those who can see the 'treasure' of life, living each day becomes a source of continual wonder and beauty.

The Nature of Things

In the past a great deal of science has been so predetermined and preoccupied with its own image and ivory tower systematics that it neglected to provide means by which ordinary people might understand and be able to relate to its existence and the knowledge of the environment it studied.

Indeed, the very strictness of its well-founded regimes and the critical analysis applied to each and every member, from professor to student, dictated the stringent manner by which everybody was scrutineered. Today, there is a lot more emphasis on public education.

Protected marine parks enable wild life to become accustomed to divers (GBR)

Once the mysteries of marine life identification are revealed, everybody gets to enjoy their eco-experience and share it with others (GBR)

Anthropomorphism (An-thro-po-mor-phism)

Dictionary definition:

"The attribution of human form and character to God, or a god, or of human character to things in nature."

I must admit I've been somewhat amused by this down through the years, especially as many of my earlier books were edited by well-meaning scientists. In all their religious righteousness they thought they were refereeing scientific papers and by the liberal use of red pen relegated my elementary attempts at communication to the scrapheap of excommunication, with this one single word.

ANTHROPOMORPHISM! (The worst of scientific sins?)

By definition, **anthropomorphism** applies equally to God, a god, or nature. Can it be that science does not recognise any relationship between humans and nature? Surely not! This is a religious ideology? Is science, indeed, a religion?

One can understand the scientific community accepting the theory of evolution and as such not believing in applying any human attributes to God, or gods...but nature? Could evolution and creation theories be just opposites of the same ideology? Could it be that science defines nature and God as the same? (Wouldn't that be wonderful?) Certainly the subject is worthy of further discussion.

In the early 1970s Reg Lipson and myself pioneered some of the first marine life identification courses ever held for divers, students and teachers in the southern hemisphere

The reality of things

In reality, the only way by which humans will ever understand nature, is if they can relate to, and recognise its existence. The best way to communicate the existence of nature is to relate it to human nature. Anybody, teacher, instructor, student, or parent can be a relater of facts and figures presenting an assemblage of caged information. Yet only those who can impart the relationships between the facts and figures and the reasons for those relationships, will ever encourage an understanding of their subject.

Bridging the gap

In order to communicate underwater nature, we must create ways to alter its present perception. We must provide the means of changing the unreality, to reality. We must bring about an understanding of what is known already and use that knowledge to bridge the communication gap between the nature of land, and the nature of water. **The only way to accomplish this is to relate the nature of land and the nature of water, to human nature.**

KINGDOM MONERA

Blue-green Algae

Phylum: Cyanobacteria (Cyan-o-bac-ter-ia)

More easily seen as a dark-coloured scum on rocky shoreline substrates than underwater, blue-green algae are extremely slippery to touch and even more to walk on.

Blue-green algae are related to the oldest recognised life forms on Earth, the Shark Bay (Western Australia) stromatolites. They produce oxygen from the water using photosynthesis and are similar more to bacteria than algae.

Recognised as the oldest recognised life forms on earth, living stromatolites are only known from Shark Bay, WA

Below the surface in the shallow sun-heated water the true nature of these living fossils is revealed

KINGDOM PLANTAE

Marine plants - Algae (The slime of your life)

Plants are the primary producers of all life on earth.

As dormant and sessile as many appear, they are without doubt the most extraordinary of beings with abilities and lifestyles that when recognised and understood, are beyond the wildest imaginations of science fiction enthusiasts.

Every life form that ever lived on this planet owes its very existence to the presence of plants - without them, life as we understand it, would neither have evolved nor progressed.

Photosynthetic wonders

All plants contain **chlorophyll** in their cells. **Chlorophyll** is a green pigment that is the essential ingredient by which plants are able to convert the energy of **sunlight, carbon dioxide** and **water** into **carbohydrates** (sugars and starches which are rich in energy) a process called **photosynthesis** (pho-to-syn-the-sis). A by-product of the **photosynthetic** process is **oxygen**, the main component gas which provides a life-support system for the entire **animal kingdom**.

There are more than 9000 species of algae in world oceans. The largest and most dominant forms occur in temperate waters, ranging in size from filamentous turfs only millimetres in height to giant kelps over 30 metres in length that can grow at a rate of half a metre a day.

Tropical reefs may support around 1000 species of algae, yet many still remain undescribed. Whereas land plants have leaves that extract carbon dioxide from the air and roots that absorb moisture and minerals from the soil, the algae have no roots in this sense, only holdfasts to anchor them to the bottom. To grow, algae need light (similar to land plants). In general terms, the subdivision of algae is based on colour: (1) **green algae**; (2) **brown algae**; (3) **red algae**.

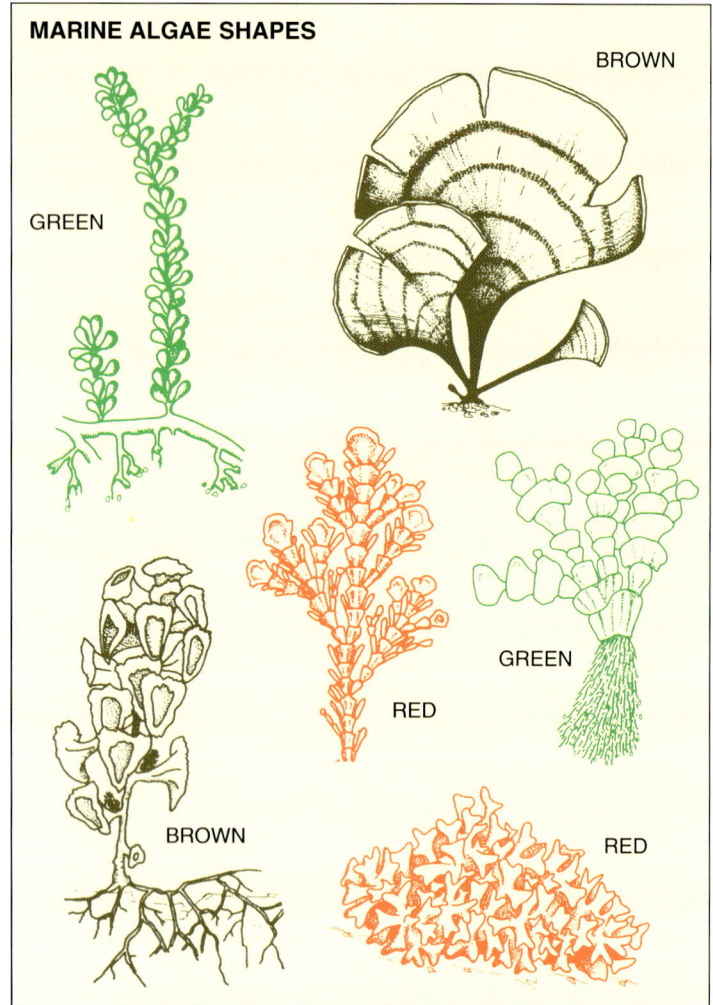

MARINE ALGAE SHAPES

Green Algae

Phylum: Chlorophyta (Chlor-o-phy-ta)

Green algae comes in an amazing range of forms, from large globular, single celled ones to those with calcified plate-like leaves. Colour is mostly green through to black. Green alga are generally found in shallow well-lit waters and once identified by a trained botanist many distinctive species can be visually identified and illustrated by photography.

Sea Lettuce

The green alga *Ulva* undergoes a cyclic alternation and can be spore-producing plant or a sexual- producing plant. Though each of these plants may look identical, they can only be identified by a specialist with a microscope. The spore-producing plant produces free swimming zoospores which swim to the bottom and attach, each growing into a complete sexual plant (male or female). When mature, these sexual plants release either sperm or eggs into the water. The fertilised eggs hatch into zygotes which settle to the bottom and grow into spore-producing plants to complete the cycle.

THE ASEXUAL - SEXUAL REPRODUCTION OF THE GREEN ALGA (*ULVA*)

SEA LETTUCE

Zoospores

Eggs — sperm

female plant   Male plant

Fertilization

Zygote

Sea lettuce Ulva sp. (green) (LHI)

Grape weed Caulerpa racemosa (green) (PNG)

Ornate Turbinaria Turbinarea ornata (brown)

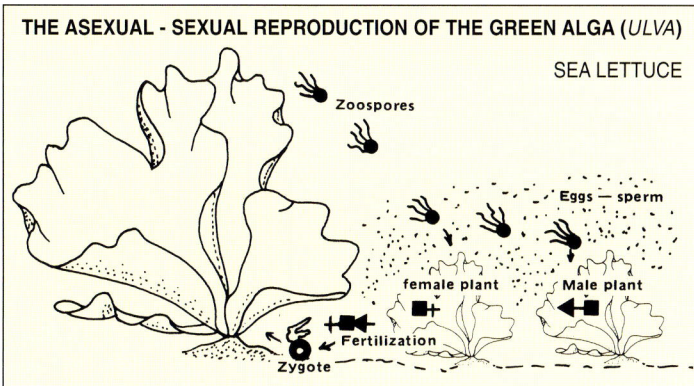

Sargassum weed Sargassum sp. (brown)

Brown Algae

Phylum: Phaeophyta (Phae-o-phy-ta)

There are thought to be over 1500 species of brown algae which mostly occur in temperate waters. What brown algae lack in diversity in the tropics they make up for in abundance, covering entire areas of the reefs. Although the phylum name comes from the Greek word *phaios* meaning "brown" the actual colour is due to the brown pigment fucoxanthin which varies from brown to yellow, blue and black. Some browns are easy to identify visually while others still require a specimen and a trained botanist.

Red Algae

Phylum: Rhodophyta (Rho-do-phy-ta)

With over 4000 species known from tropical reefs, red alga are the most diverse in shape, pattern and design and are found from the intertidal area down to at least 100 metres. They range from soft spongy textures to upright clumps of pink brittle skeletons with articulating joints. Coralline encrusting red algae covers the reef rims, cementing living and dead corals together to form protective ramparts against the ever-erosive ocean.

Although some species can be visually identified, many others remain a challenge and require a specimen and experienced botanist to establish a name.

Fungi-form Peyssonelia Peyssonelia sp (red) Red algae of this type are very common on walls

Seagrasses

Phylum: Magnoliophyta (Mag-no-lio-phy-ta) Angiosperms

Generally found on muddy or sandy sea floor in sheltered areas, in bays and lagoons or in clear, deeper waters. There are 55 species of seagrasses throughout the world and although commonly referred to "grass" are, in reality, true flowering plants that have adapted to the marine environment.

Seagrass meadows are a dynamic part of the eco-system and, as such, should be protected from human destruction. Seagrasses have roots and flowers and like all plants require sunlight for **photosynthesis**. By doing so, they are able to increase the oxygen and reduce carbon dioxide and acidity in the water around them.

Identification to genus on some can be made by visual means but species definition requires a botanist and specimens.

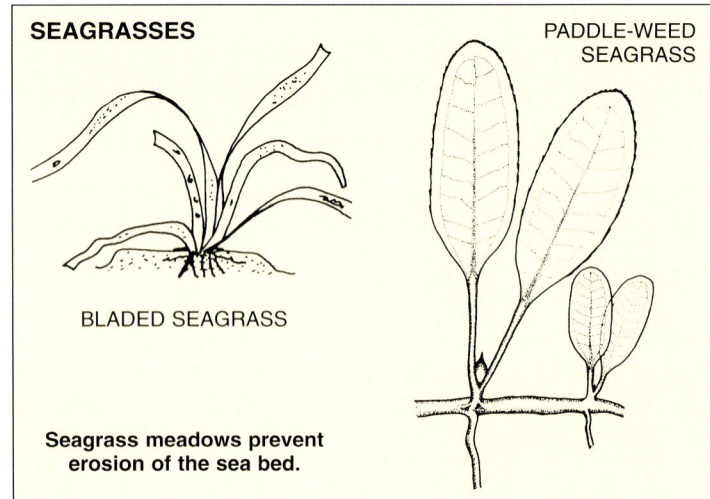

SEAGRASSES

PADDLE-WEED SEAGRASS

BLADED SEAGRASS

Seagrass meadows prevent erosion of the sea bed.

Tropical seagrass Enhalus acorides grows tall in the shallows (PNG)

Seagrass scientist checking strap-weed seagrass Posidonia sp. (Albany WA)

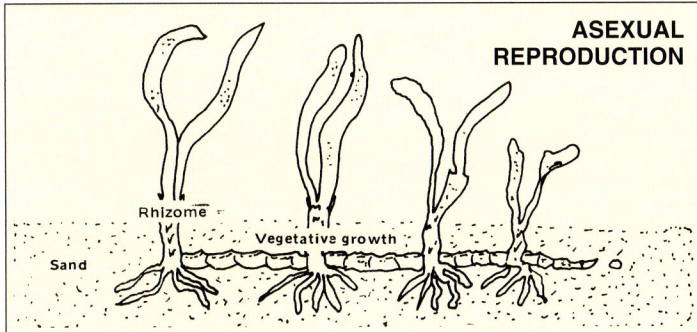

ASEXUAL REPRODUCTION

Rhizome

Vegetative growth

Sand

MANGROVES

GREY

RED

Mangroves

Strictly speaking, mangroves are not marine plants in total; neither are they (in total) land plants. Mangroves are unique, for they inhabit the meeting place of land and sea. Twice a day their root systems are completely immersed in salt water in which no other land plant could exist and twice a day the same root systems are exposed to long periods in air.

No other community of flowering plants is subjected to such a range of environmental extremes as is produced by conditions in the intertidal zone. Severe oxygen depletion in the mud, high salinity, low salinity (fresh water flooding), temperature fluctuations, nutrient rises and declines, open ocean, to semi-closed flushing systems.

Mangroves are similar to other land plants whereby they extract oxygen and carbon dioxide from the air. However, unlike other land plants they breathe through minute pores in their trunks, stems, branches and roots. Many species have evolved specialised root systems to enable respiration, e.g. the peg-like erect roots growing up through the ground around mangrove trees are called **pneumatophores** (new-mat-o-phore). There are buttress roots, stilt roots, knee roots, and aerial roots, all adapted to breathing.

The excess salt taken up by the trees during salt water flooding is excreted through the leaves of some species, while others, can take in the salt water, and filter out the salt. In other functions, most mangroves are in common with other land plants, with reproduction by way of flowers, pollinated by bees, butterflies and moths. Some, however, are **viviparous** and produce seedlings with partly-developed root systems while they are still attached to the parent tree.

The grey mangrove Avicennia marina has peg-like root systems called pneumatophores (Qld)

Red mangroves have stilt-like roots. (Loloata Isl. PNG)

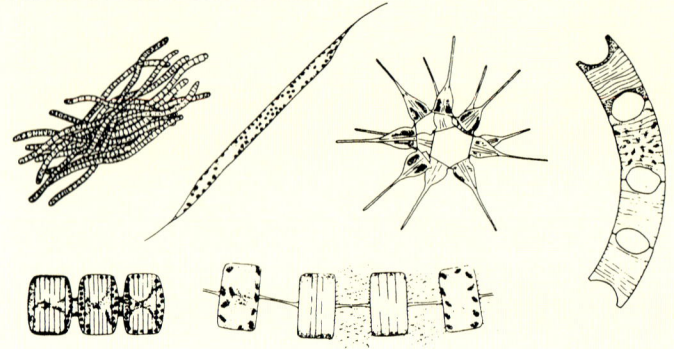

PHYTOPLANKTON FORMS

Phytoplankton (Phy-to-plank-ton)

Although Danish botanist Anders Sandoe Ørsted did not invent the plankton net, he was the first to discover the presence of **microalgae**, or **phytoplankton**. Phytoplankton are the basis for all life in the sea. Ørsted realised this and in 1847 wrote a paper on their importance as the ocean's primary producers.

Human attitudes towards new concepts were no different 140 years ago from what they are today, and Ørsted's treatise on his discoveries was ignored equally in his home country and by scientific colleagues throughout the world. It was not until some 50 years later, at the turn of the century, that Ørsted's studies were accepted, and today they are honoured as among the 10 most important scientific works of the nineteenth century.

Phytoplankton are, for the most, minute, **single-celled algae** which inhabit the sunlit surface layers of the ocean down to 200 metres in clear water. These microscopic plants form the food base that in one way or another, **supports all life in the seas.** A single litre of sea water may contain as many as 10 to 20,000,000 individuals and represent a number of species. Similar to all plant life, **phytoplankton** require sunlight, water, carbon dioxide and nutrients to grow and **photosynthesise**. A by-product of this **photosynthetic** process is oxygen and it is calculated that **phytoplankton** and marine plants produce up to 60 per cent of the earth's **oxygen**. Some of the most well-known classes are known as **diatoms** and **dinoflagellates** of which there are over 7000 species recognised in the Indo-Pacific.

Not all **phytoplankton** are free living. The corals of the Great Barrier Reef contain in their living tissues the greatest assemblage of **sedentary symbiotic dinoflagellate** communities in the world, the **zooxanthellae**. While the **zooxanthellae** provide beneficial life-progressing relationships with many different groups of sea creatures such as clams, sea squirts, sea anemones, corals, flatworms and nudibranchs, some species of free-living **dinoflagellates** are (when in bloom) toxic to other sea creatures, and humans. These blooms of diatoms and dinoflagellates are commonly known as **red tides** and often coincide with high increases in **nutrients** such as **nitrates** and **phosphates** in a localised area.

Large populations and agricultural centres near rivers discharge huge amounts of industrial (heavy metals, toxic chemicals), domestic (sewerage) and agricultural wastes (cane farms) into the sea, raising the **nutrient** levels and, in some cases, this practice is responsible for causing **red tide** outbreaks, resulting in huge fish kills and **toxic properties** in oysters, mussels and scallops.

In tropical waters, from September on during summer, blue-green alga *Trichodesmium erythraeum* (sea sawdust) die in large numbers and float to the surface to cover the waters with gigantic red-brown "rafts". This is probably the only time divers see **phytoplankton** in the sea, though they swallow hundreds of thousands of them every time they get a mouthful of salt water.

Sea sawdust Trichodesmium erythraeum covers huge areas of ocean during summer (GBR)

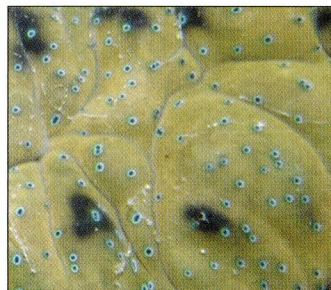
Clam mantle (Tridacna) patterns are formed by the zooxanthellae in their tissues

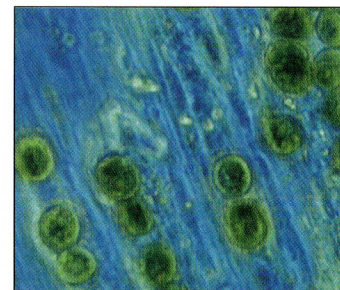
Microscopic zooxanthellae in clam tissue

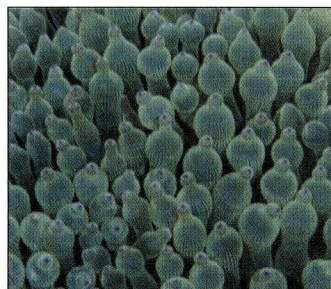
Sea anemone with zooxanthellae (LHI)

Sea anemone without zooxanthellae (Qld)

KINGDOM PROTISTA

Protozoans (Pro-toe-zo-ans)

Protozoans, the "first animals" as their name suggests, are the most **prolific animals in our seas**. Besides the free living species, every known animal plays host to more species of **protozoans**. There are even protozoans which are inhabited by smaller protozoans. These microscopic animals are called ciliates, amoebas, dinoflagellates, **forams** and radiolarians and occur throughout every aquatic environment.

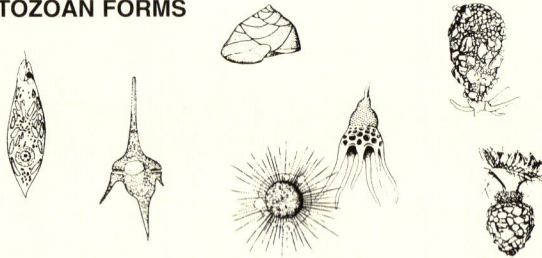

PROTOZOAN FORMS

The illustrations are redrawn from Herring and Clarke and Davis (early 1990s)

Cauliflower foram Sporadotrema sp. (PNG)

White spiky foram Sporadotrema sp. (PNG)

Vertebra forams Marginopora vertebralis cover the bottom at some locations (PNG)

Foraminifera

Phylum: Sarcomastigophora (Sar-co-masti-go-fora)

Between the kingdoms of plants and animals there are transitional creatures which often maintain the features of both. One such group are the forams of which there are many worldwide species. Twenty or so can be found on tropical reefs and a few cold water species have bipolar distribution.

There appear to be three main groups - (1) those that are sessile and attached to the bottom, beneath rocks or under ledges and caves; (2) those that have free living benthic lifestyles, and (3) those that are pelagic.

Forams are actually **single-celled amoeba-like protozoans** which often have complex calcium carbonate shells containing jelly-like protoplasm. Some forams have simple circular body forms (*Marginopora*) coated in protoplasm strands which enable them to trap prey. Sedentary forms live in the dark beneath rocks and although pigmented their lifestyles are little known. However, it appears that they trap prey with sticky protoplasm which is strung out on skeletonal projections.

Some of the structures are very dense in their construction and without the benefit of **symbiotic algae** it may take many years to lay down their skeleton. On some sand cays and beaches almost the entire content is made up of foram skeletonal material from margin or star sand forams.

Forams reproduce sexually and asexually (splitting) and in some species the **symbiotic algae** is actually passed along to the offspring during division of the cells; the protoplasm in the parent body is divided equally amongst the offspring and the parent ceases to be.

Once identified from a specimen, visual identification of calcified species is possible from a close up photograph.

KINGDOM ANIMALIA

Sponges

Phylum: Porifera (Pore-if-era) pore bearers

Sponges are considered by some to be the **first multicellular animals;** their fossil remains have been traced back to some 650 million years ago. These simple primitive life forms are widespread throughout temperate and tropical seas with around 15,000 species worldwide.

Sponges are **sedentary individual animals** (not colonies) covered by a thin "skin" which houses multitudes of roving cells. In general, the **minute pores** seen all over a sponge are the **inhalant pores** where water is sucked in to breathe and feed by filtering out plankton and suspended sediment. **The larger holes (oscula)** are the **exhalant openings** where waste water and products are expelled.

Sponges range from minute encrusting species under rocks, to massive structures one and a half metres (five feet) tall. Their shapes can be very variable and they come in almost every colour of the rainbow.

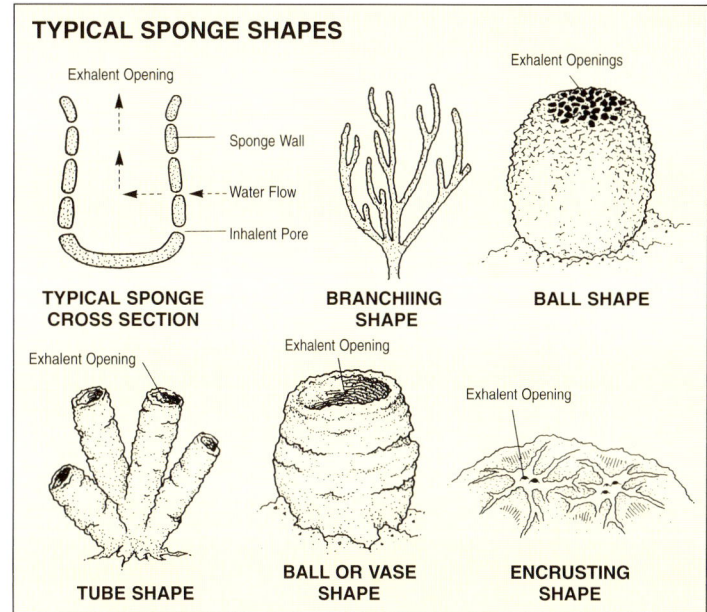

TYPICAL SPONGE SHAPES

TYPICAL SPONGE CROSS SECTION — Exhalent Opening, Sponge Wall, Water Flow, Inhalent Pore

BRANCHIING SHAPE

BALL SHAPE — Exhalent Openings

TUBE SHAPE — Exhalent Opening

BALL OR VASE SHAPE — Exhalent Opening

ENCRUSTING SHAPE — Exhalent Opening

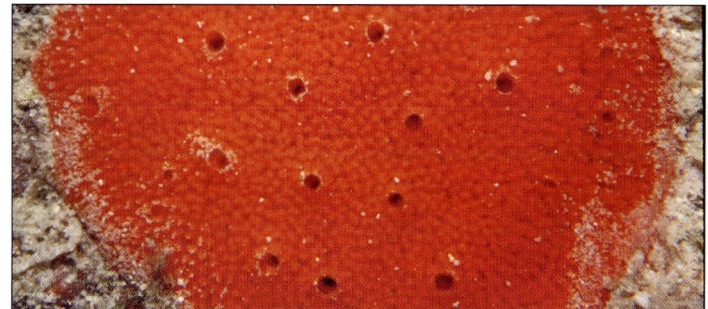

Encrusting growth form of sculptured sponge Pseudaxinella cf australis. The larger exhalant openings are fairly evenly spaced (Mal.)

In the past, sponge identification has been very difficult due to their **taxonomy** being based on **minute spicules** in the body of the skeleton. However, taxonomists at the Queensland Museum are reclassifying thousands of species with modern studies cross-referenced with living pictures.

There are two major classes of sponges.

Class: **Demospongiae** (Demo-spong-iae) siliceous sponges

Siliceous sponges have **fibrous skeletons** made up of tough keratin-like material called **spongin** which is very resistant to decomposition and wash up on beaches. The **spicules** of **siliceous sponges** are made of **silica**.

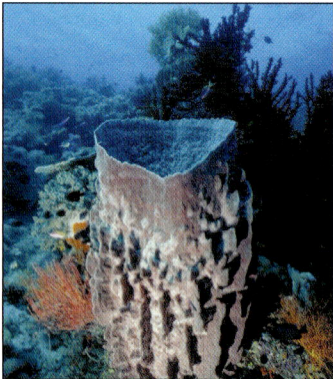
Stony sponge *Astrosclera willeyana* (living fossil) (PNG)

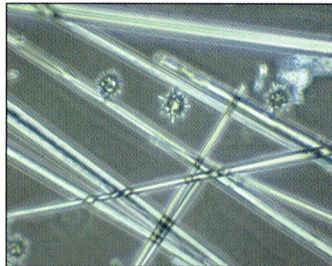
Encrusting sponge; volcano-like exhalant openings (PNG)

Class: **Calcarea** (Cal-car-ea) calcareous sponges

There are much fewer species of calcareous sponges, and in general, they are yellow, white or light brown. **Calcareous sponges** have **spicules** comprised of **lime** (calcium carbonate) and these sponges have **no spongin fibres** in their skeleton.

Barrel sponge *Xestospongia testudinaria* (PNG)

Microscope slide of sponge spicules

Siliceous sponge skeleton (PNG)

Turret sponge (unidentified AMPI/Por.454)

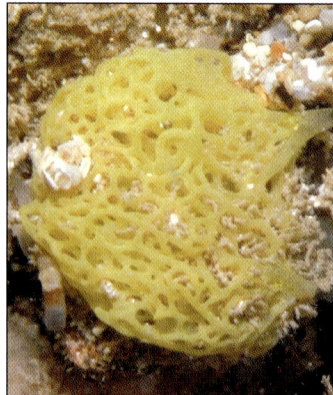
String-ball calcite sponge *Clathrina sp* (LHI)

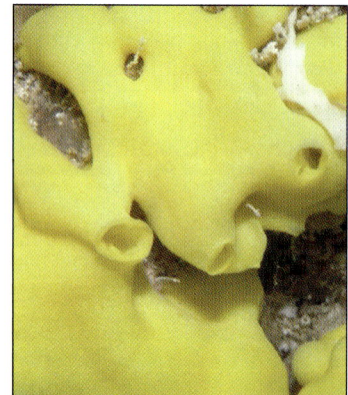
Yellow calcite sponge *Leucetta sp* (LHI)

CNIDARIANS

Phylum: Cnidaria (-nid-ar-ia) stingers

At one time all animals with **radial symmetry** and simple sac-like bodies were placed in the phylum Coelenterata. Today, scientists recognise two separate phyla at this level: **Cnidaria** and **Ctenophora**.

All the animals illustrated in this section belong to the larger of the two phylums, Cnidaria. The term **'Cnidaria' refers to the power to sting**, a feature of those animals which have **special stinging cells** in their bodies.

The basic structural unit of a typical cnidarian is a flower-like polyp. This polyp has **no breathing mechanism, no blood** and **no excretory system**. In short, it is a sac-like organism with an opening at one end surrounded by one or more circlets of tentacles. The vital functions of **respiration, excretion** and **food distribution** are achieved by **simple diffusion**.

The tentacles are hollow containing a space which connects to the gut and are armed with cells called **cnidoblasts** which house multiple numbers of stinging nematocysts. Once the prey is subdued, the tentacles manoeuvre it to the mouth. It then passes to the stomach where it is digested and the useful products are absorbed. The **refuse** is **regurgitated** and **ejected** via the **mouth**.

There are about **9000 species of cnidarians living in the world's oceans**. (This number is only an estimate due to continuing changes in taxonomy and new discoveries.) The life histories of cnidarians are often varied and complex. The following is only a general account. Within the phylum cnidaria, there are two main body forms - **free swimming medusae** (sea jellies) and **sedentary polyps** (hydroids, corals and sea-anemones). Both body forms are radially symmetrical, with the mouth located at the centre. The basic differences between the **free-swimming medusa** and the **sedentary polyp** are their **orientation** and their **mode of life**. Most **medusa swim** with the

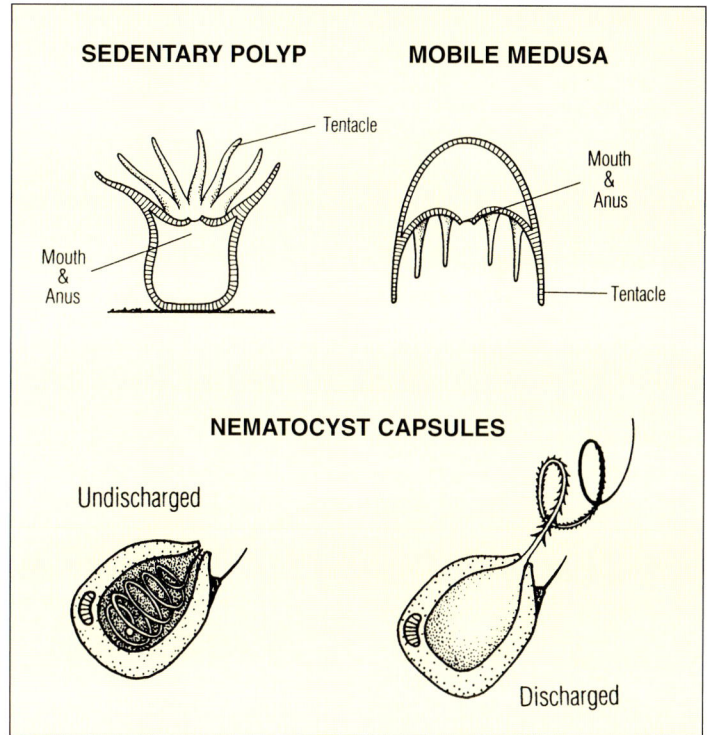

SEDENTARY POLYP — Tentacle — Mouth & Anus

MOBILE MEDUSA — Mouth & Anus — Tentacle

NEMATOCYST CAPSULES

Undischarged

Discharged

mouth and tentacles facing **downwards**, while the **stationary polyp** is attached to a substrate with the mouth and tentacles facing **upwards**, outwards, or downwards. One important feature of Cnidarians is their ability to form colonies.

Although all cnidarians possess nematocysts, only a few have the capacity to harm humans. Some hydroids, fire corals, sea jellies and sea anemones can injure humans and a few, including the tropical box jellies (Cubozoa) have caused the deaths of many swimmers.

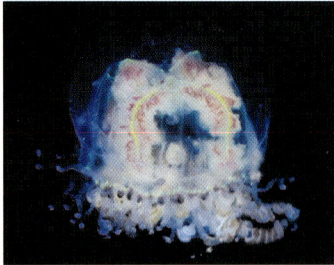
Hydroid (intermediate stage) Limnomedusian Olindias phosphorica (stings) (Qld)

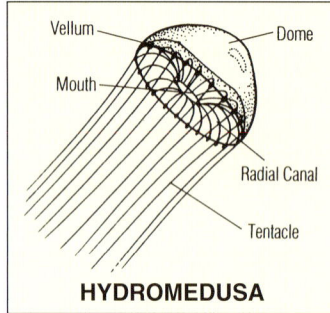

HYDROMEDUSA

Class: **Hydrozoa** (Hyd-ro-zoa) sea ferns, hydrocorals, fire corals, Portuguese man-of-war

The order Hydroida (sea ferns) is one of the few cnidarian groups that is more diverse in temperate waters than in tropical seas. Although very common, hydroids are certainly not familiar to most divers. The majority are low-profile clusters of fine, **fern-like structures** that tend to blend in rather than stand out. Many species are very small, and may live on other organisms, such as seaweeds, sponges and shells. However, those in tropical waters such as *Agliophena* and *Macrorhynchia* grow in large clumps and sting virulently. By contrast, the colonial hydrocorals and **fire corals** have **hard, calcified structures**. Fire corals may have massive colonies and their powerful nematocysts can inflict severe stings on humans.

Philippine hydroid Macrorhynchus philippinus (stings) (GBR)

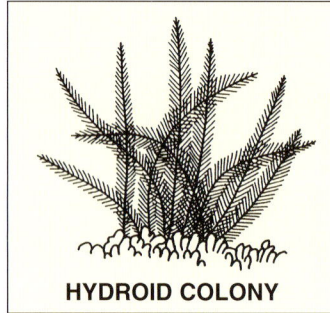

HYDROID COLONY

Solitary polyp of Ralpharia sp. shows reproductive stages ready for release

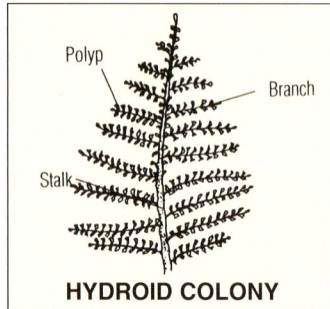
Fire coral tentacles Millepora tenera (stings) (PNG)

Hydroid polyps extended for feeding (PNG)

HYDROID COLONY

FIRE CORAL POLYPS (microscopic view)

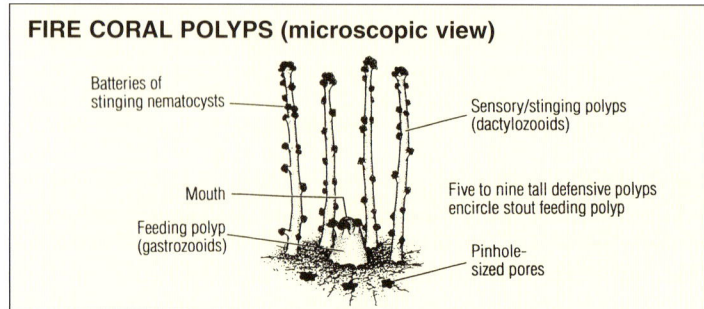

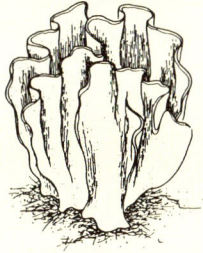

FLAT SIDED FIRE 'CORAL'

Flat-sided fire coral Millepora platyphylla
in protected habitat (Indo.)

Yellow hydrocoral Distichopora sp. (PNG)

Fire coral Millepora 'platyphylla' on reef rim

Fire coral 'head' Millepora sp. (PNG)

Purple hydrocoral Distichopora sp. (GBR)

Red hydrocoral Distichopora sp. (Van.)

Elegant hydrocoral Stylaster elegans (GBR)

Lord Howe hydrocoral Stylaster sp.

Pink hydrocoral Distichopora sp. (Maldives.)

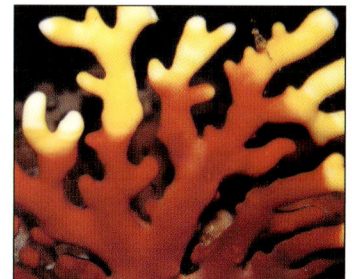

Yellow & red hydrocoral Distichopora sp. (PNG)

Stinging siphonphore Physophora hydrostatica (photo: Rudie Kuiter)

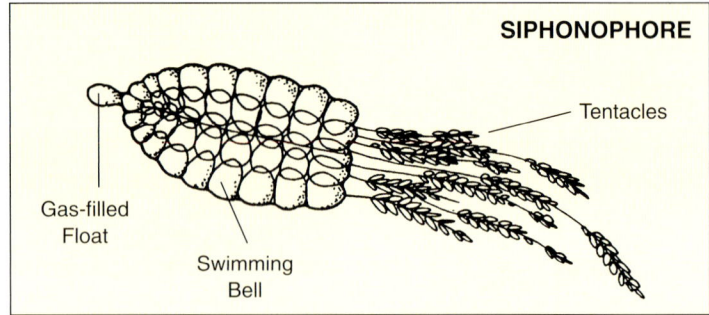

Portuguese man-o-war Physalia physalis (stings) (NSW)

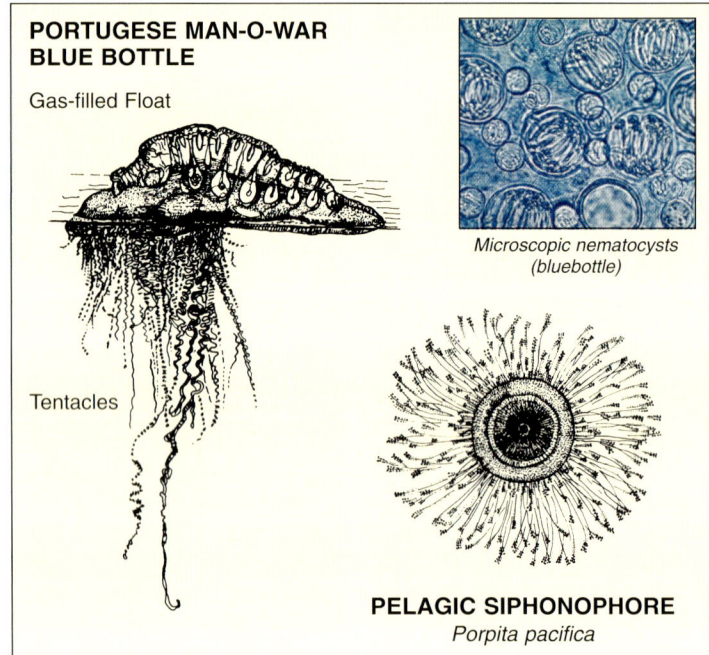

Feeding zooids on underside

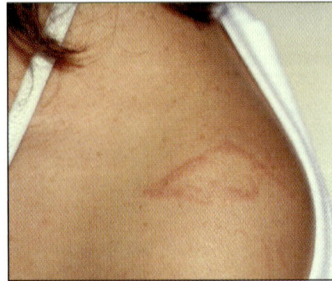

Portuguese man-o-war sting (Qld)

SIPHONOPHORE

Tentacles

Gas-filled
Float

Swimming
Bell

**PORTUGESE MAN-O-WAR
BLUE BOTTLE**

Gas-filled Float

Tentacles

*Microscopic nematocysts
(bluebottle)*

PELAGIC SIPHONOPHORE
Porpita pacifica

Class: Scyphozoa (Scy-pho-zoa) sea jellies

Sea jellies belonging to the class Scyphozoa are defined as cnidarians where the medusa phase is dominant and there is a subordinate polyp stage. The anatomy of sea jellies is quite complex with the extensive mesogloea contributing to their gelatinous bulk. Transfer of food from the digestive areas to the tissues is helped by a system of canals; balance organs assist in swimming. **Pulses of contraction** sweep over the **muscles of the bell** and cause the tissues to make **characteristic beats** which enable the **animal to swim** through the water.

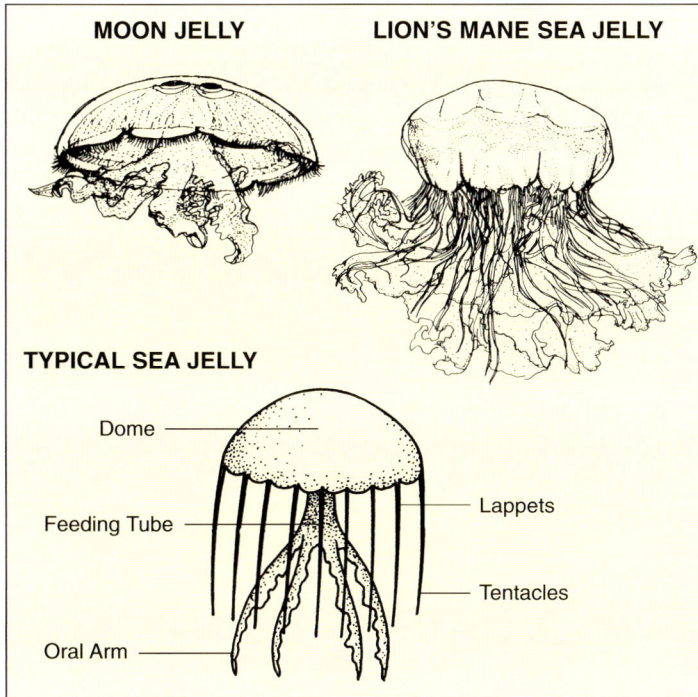

Mosaic sea jelly Catostylus mosaicus (Photo: Nigel Marsh) (Qld)

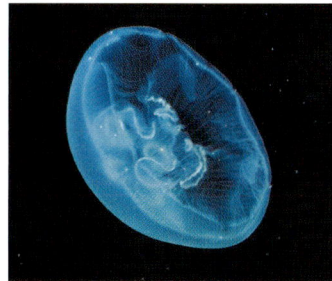

MOON JELLY

LION'S MANE SEA JELLY

TYPICAL SEA JELLY

Dome

Feeding Tube

Oral Arm

Lappets

Tentacles

Moon jelly Aurelia aurita (NSW)

Upside-down sea jelly Cassiopeia andromeda (PNG)

Papuan sea jelly Mastigias papua (PNG)

Lion's mane sea jelly Cyanea capillata (Photo: Nigel Marsh)

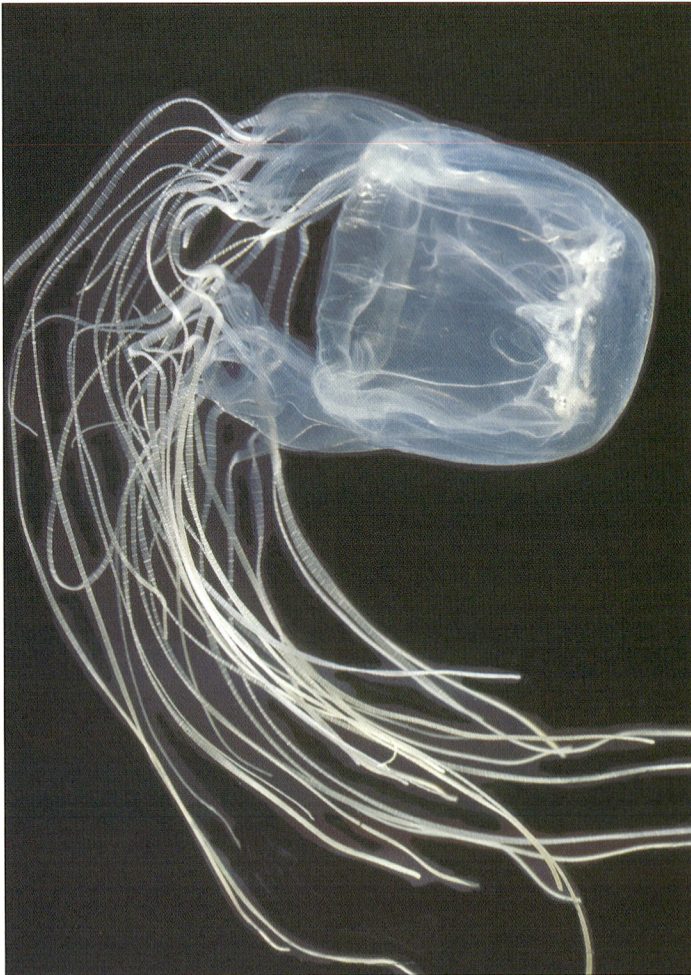

Box jelly Chironex fleckeri (deadly) box jellies have killed over 70 Europeans in Australia (NT)

Class: Cubozoa (Cu-bo-zoa) box jellies

Box jellies are amongst the **most venomous animals** in the world. Although box jellies are tropical in their distribution, the largest and most dangerous are in the waters of the southern hemisphere. The common term, "box jelly", describes this group admirably. The body has a tough gelatinous composition, a box-like, four-sided bell with one or more tentacles attached to each corner. At the underside of the bell the edges curve under, forming a "skirt" yet not enclosing the body area. Sense organs made up of a balance mechanism and an eye are situated on each of the four sides, usually on the central perpendicular axis towards the base of the bell. Under the bell, hanging from the top, is the stomach and tubular mouth.

The box jellies are made up of two families: members of Corybdeidae have **four single tentacles** and are **mild stingers**, while the Chirodropidae have **four clusters of tentacles**, with some species being **deadly to humans**.

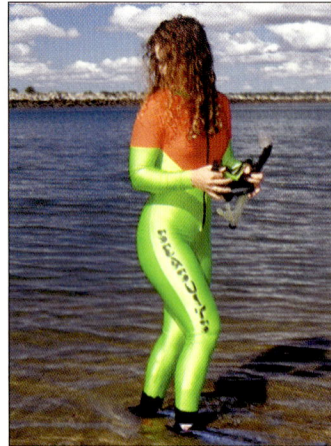

Lycra suits offer excellent protection (Qld)

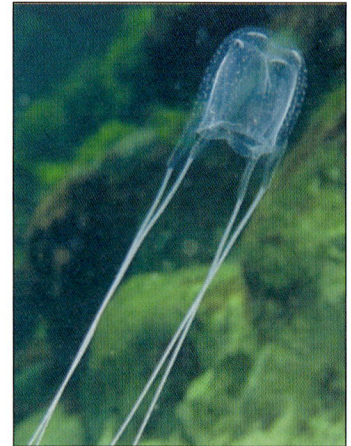

Jimble Carybdea rastoni (SA)

Class: Alcyonaria (Alcy-on-ar-ia) soft corals, sea fans, sea whips, sea pens

All these types of sedentary colonial organisms belonging to the class Alcyonaria can be easily distinguished from other members of the cnidarian phylum as their **polyps have eight tentacles (Octocorals) and the tentacles are always fringed, or pinnate.**

Soft corals

Unlike the **stony corals**, most **soft corals** have **no true skeleton** as such. Their soft tissue-like bodies are flexible and various species may be soft and squelchy (*Xenia*) or firm and leathery (*Sarcophyton*) while others (*Dendronephthya*) hold themselves erect by pumping water into their interconnecting body cavities. All species, no matter how soft or pliable, contain **minute spicules** of **calcium** which in many cases help to strengthen the body walls. It is these **spicules** which scientists use to describe different species of alcyonarians, and in the genus *Dendronephthya* some types of spicules can be seen in the body walls.

Being cnidarians, soft corals have nematocysts in their polyps which sting and catch planktonic prey but these do not affect humans. However, some of the spiky soft corals (*Dendronephthya*) have protective spikes arranged around their polyp clumps (which act in a similar fashion to the thorns on rose bushes) by helping to protect the soft coral from browsing nudibranchs or ovulid cowry animals.

Soft corals also have extremely potent chemical defences and some even have chemicals with which they can attack and kill other organisms whose territory they wish to invade. Many of these **substances** are being investigated by **bio-chemists** to see if they have properties which will **inhibit cancer cells**, or other **diseases in humans**. Some shallow water dwelling **soft corals** (those in the light) have **symbiotic zooxanthellae** living in their tissues.

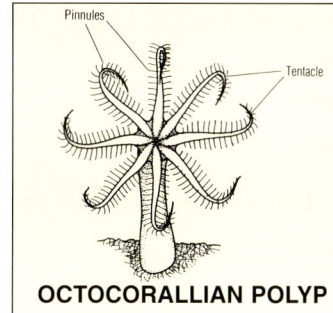

OCTOCORALLIAN POLYP

Soft coral polyps Carijoa sp. (PNG)

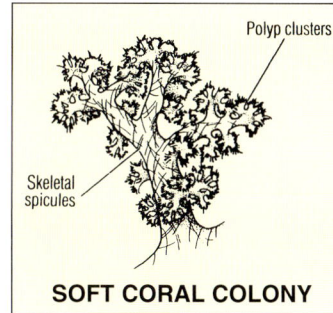

SOFT CORAL COLONY

Spiky soft coral Dendronephthya sp. (PNG)

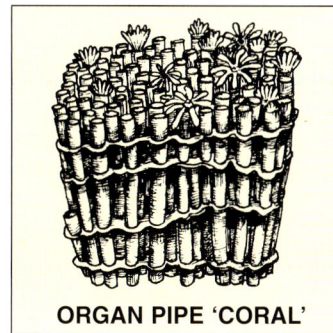

ORGAN PIPE 'CORAL'

Organ-pipe coral Tubipora sp. (PNG)

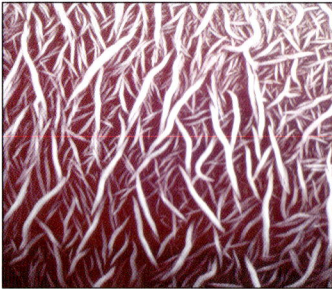
Skeletal spicules Dendronephthya sp. (PNG)

DENDRONEPHTHYA SPICULES

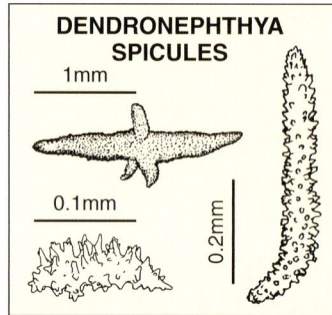

1mm

0.1mm

0.2mm

Soft coral polyp Clavularia sp. (PNG)

Leather coral Sarcophyton sp. (GBR)

Spiky soft coral Dendronephthya sp. (colony)

Dead-man's fingers soft coral Sinularia sp.

Gorgonian sea fans and sea whips

As alcyonarians, gorgonian sea fans and sea whips have eight-tentacled fringed polyps and on deeper parts of the reefs these polyps can be seen out feeding during the day, especially in areas of strong current, or on the incoming tide.

Most gorgonians have strong **wire-like central skeletons** and are very flexible, others may be very brittle. Tropical waters have more species of gorgonians and soft corals than temperate waters, though in both areas these animals exhibit extremely brilliant colours and intricate growth patterns. Gorgonians secrete a **special skeletal protein** called **gorgonin** which is embedded with spicules and quite hard to the touch. Many colonies orientate to build their fan at right angles to the prevailing currents. Some species found in deep waters or in some lagoons may reach a height of three metres and be four metres across. On some deep water ridges on coral sea reefs, the gardens of gorgonian fans run like giant netted fences along the tops of the ridges for as far as the eye can see down from 50 metres.

INTERCONNECTED NET-LIKE BRANCHING

PINNATE BRANCHING

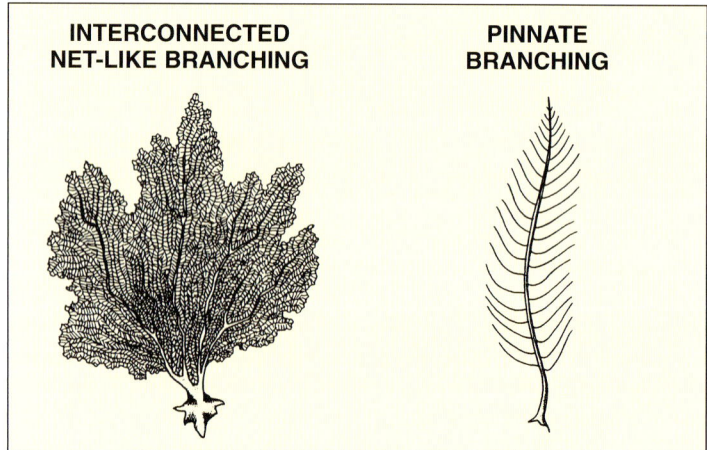

DICHOTOMOUS BRANCHING

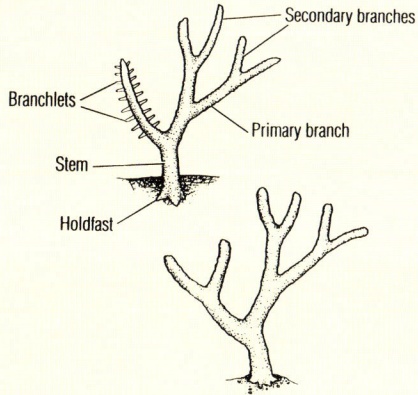

Secondary branches

Branchlets

Primary branch

Stem

Holdfast

LATERAL BRANCHING

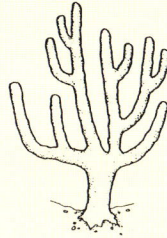

GORGONIAN BRANCH (showing calyx and aperture variations)

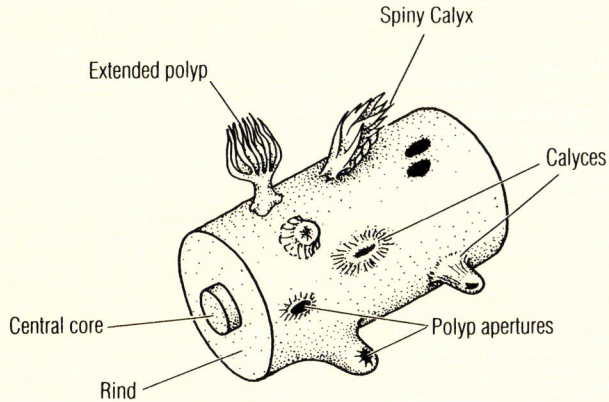

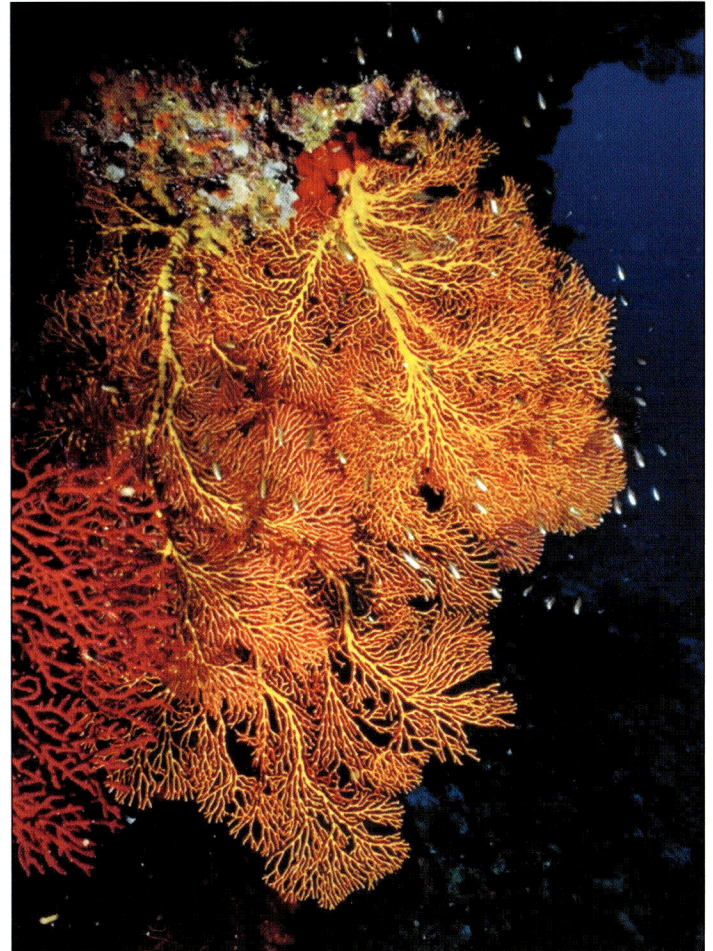

Spiny Calyx

Extended polyp

Calyces

Central core

Polyp apertures

Rind

Large gorgonian sea fans (Melathaea sp.) live below 20 metres on many tropical Asia/Indo-Pacific reefs (GBR)

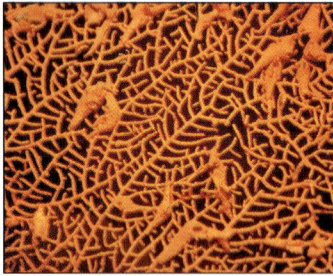
Gorgonian sea fan with commensal barnacles (PNG)

Whip-branched gorgonian Ctenocella sp. (PNG)

Sea pens

Sea pens are quill-shaped colonies of **octocorals** which live in **soft bottom** (sand or mud) areas where there are moderate currents (channel bottoms). One specialised polyp is modified to support the body. Most have an internal skeleton and they are able to expand and retract down into the sand at will.

Although common in some areas of tropical waters they appear to be sensitive to light and divers tend to see more at night. Some species can produce **bioluminescent** light when disturbed or touched.

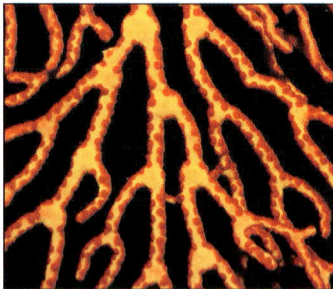
Yellow-red sea fan Melithaea sp. (Fiji)

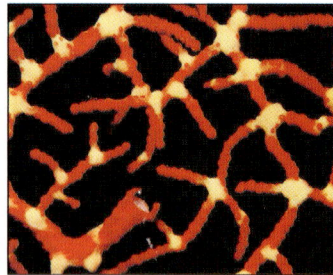
Red and yellow Clathraria Clathraria sp. (Maldives)

Fat sea pen Cavernularia obesa (PNG)

Fat sea pen polyps Cavernularia obesa (PNG)

Gorgonian sea whips Junceella sp. (Qld)

Gorgonian sea whip polyps Junceella sp. (PNG)

Feathery sea pen Virgularia sp. front (PNG)

Feathery sea pen Virgularia sp. back (PNG)

Class: Anthozoa (An-tho-zo-a) sea anemones

Well represented throughout the Asia/Indo-Pacific tropical and temperate areas, anemones vary in size from the giant tropical forms (*Heteractis* and *Stichodactyla*) which may reach a size of one metre across, down to the small diaphanous anemones only 12mm across.

Similar to stony corals many tropical anemones contain **microscopic zooxanthellae** alga in their tissues giving them a green or brown colouration. Some heave extraordinary shapes and patterns and a few tropical species (*Dofleina* and *Actinodendron*) are **virulent stingers** causing **humans intense pain** and tissue damage on contact with bare skin.

While most of the tropical forms are well known due to some providing safe havens to commensal clown and anemonefish, the temperate species are not at all well known in the scientific literature, many remain undescribed as new species without names. Besides the well publicised partnerships between anemones and anemonefish, divers can find a number of species of commensal shrimps, porcellanid crabs, portunid crabs and even brittle stars living in sea anemones.

STINGING SEA ANEMONE
ACTINODENDRON PLUMOSUM
(day)

SEA ANEMONE
ACTINODENDRON PLUMOSUM
(night)

SEA ANEMONE

Tentacles
Slit-like Mouth
Oral Disk
Body

SEA ANEMONE IN REEF

Bulb-tentacle sea anemone
Entacmea quadricolor (PNG)

Magnificent sea anemone
Heteractis magnifica (GBR)

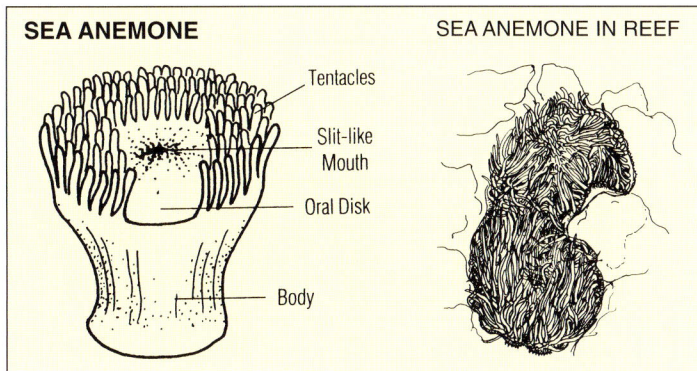

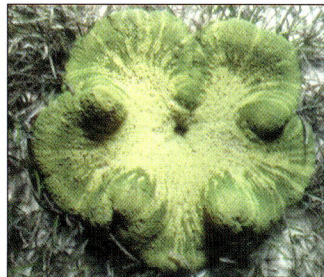

Haddon's sea anemone
Stichodactyla haddoni (Qld)

Sticky sea anemone
Cryptodendrum adhaesivum (Indo.)

47

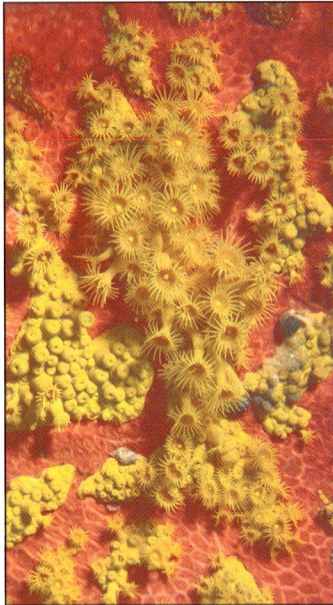
Commensal yellow zoanthid Parazoanthus
sp. (Vic)

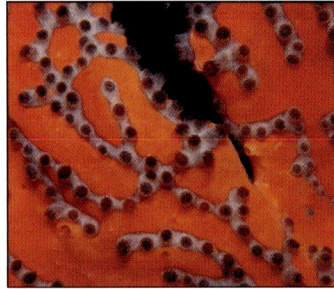
Grey commensal zoanthid Epizoanthus sp.

Stripe-disc zoanthid Protopalythoa sp. (Mal.)

Manton's zoanthid Zoanthus mantoni
(Mal.)

Encrusting zoanthid Palythoa caesia
(GBR)

Class: Anthozoa zoanthids

Zoanthids are best described as communal **coral-like animals**, **without skeletons** which live on reefs in tropical waters investing areas in extensive sheet-like formations. The best known is the intertidal species *Palythoa* which is poisonous. The Hawaiian warriors of old dipped the tips of their spears into the juices of zoanthids to render them deadly. In temperate seas, strange sausage-like specimens of the genus *Zoanthus* grow in clumps on reefs in depths down to 60 metres.

The most colourful zoanthids are those from the genus *Parazoanthus* which live commensally on and in sponges, mostly in temperate seas, causing the most intricate patterns to form on the sponge surface. In deeper waters of southern Australia the bright yellow *Parazoanthus sp.* is one of the most prominent cnidarians. **Zoanthids** feed on **plankton** and apart from the colourful species they mostly go unnoticed by divers.

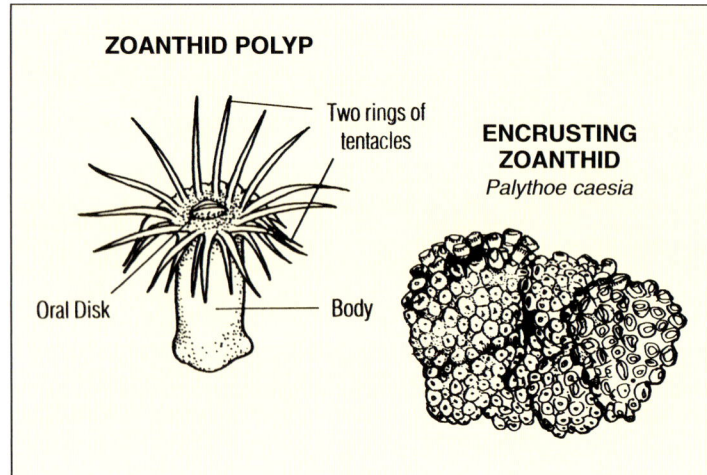
ZOANTHID POLYP

Two rings of tentacles

ENCRUSTING ZOANTHID
Palythoe caesia

Oral Disk

Body

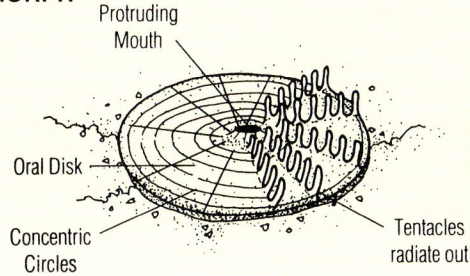

CORALLIMORPH

Protruding Mouth

Oral Disk

Concentric Circles

Tentacles radiate out

Balloon corallimorph Amplexidiscus fenestrafer GBR)

Class: **Anthozoa** corallimorphs

Looking for all the world like small beaded carpets of flat anemones, corallimorphs range in a host of brilliant colours, pinks, blues and greens. They live mostly in **large colonies** with individual polyps up to 50 to 70 mm across. However, the giant balloon corallimorph *Amplexidiscus* sp. may grow 30cm across. Most species of larger corallimorphs are tropical. The very common little jewel corallimorphs (*Corynactis*) carpet the insides of caves in southern Australia.

Carpet corallimorph Discosoma sp. (stings) (LHI)

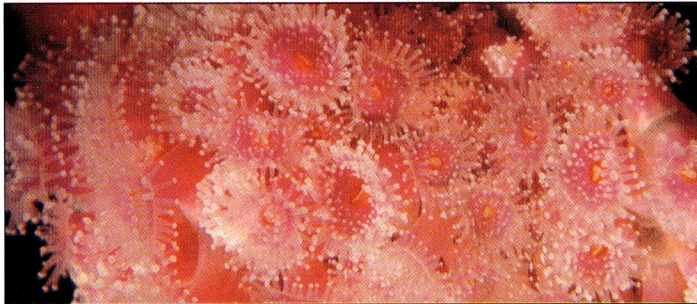

Southern corallimorph Corynactis australis (NSW)

Scalloped corallimorph Discosoma sp. (Indo.)

Fringe-mouthed corallimorph Discosoma sp. (Sey.)

Staghorn and table corals Acropora sp. often dominate on semi-protected reefs (GBR)

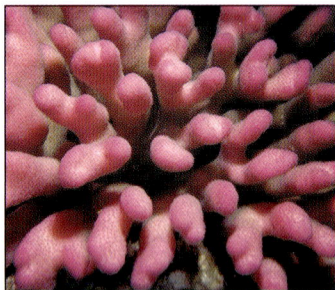
Pistillate coral Stylophora pistillata (GBR)

Hedgehog coral Galaxea fascicularis (Mal.)

Cabbage coral Turbinaria reniformis (GBR)

Brain coral Leptoria phrygia (Sol Is)

Class: Anthozoa stony corals

Although **stony corals** do exist in **temperate waters** and in very clear water may form large plate-like formations even down as deep as 65 metres, it is in **tropical seas** that the **true reef building corals** flourish. While temperate seas in Australia may support around 20 species of corals the **tropical reefs** have around **350 species.**

Some corals such as the staghorn corals (*Acropora*) may have as many as 60 different species described, though future studies may reveal some of these to be forms of the same species growing in different micro habitats. Underwater photography, collection and a visual identification system of corals was first pioneered by the author and this system has now been advanced by diving scientists who are able to identify the majority of stony corals (at least to genera) by observing a good close up photograph. However, in situ, recognition of many species remains difficult and even with specimens in hand some types continue to be ambiguous.

Many tropical coral reefs of the Asia/Indo-Pacific region are dominated by staghorn corals in their many shapes and growth forms. Warty corals (*Pocillopora*) and needle corals (*Seriatopora*) are also very common.

In the shallow water lagoons massive (often circular) flat-topped stony coral colonies may be seen at low tide. These are modified versions of the giant mound corals *Porites*. The flat-topped versions are called **micro-atolls** because the inside is only dead skeleton surrounded by living coral on the outside. These **micro-atolls** are formed by being **exposed** to **air** and **sun** for long periods at **low tide**. The height of the atoll corresponds to the low tide level in the lagoon.

In general, the more robust stony corals are found in exposed situations, or grow smaller and stouter than their counterparts in deeper water or sheltered lagoons. The polyp of a stony coral has **six tentacles** or is made up of **multiples of six** (12-18) and the **tentacles** are **simple**, not pinnate, as in the soft corals (*Alcyonarians*).

The majority of tropical stony corals are colonial, building their massive structures by **budding** or **splitting** (asexual reproduction) in a **cloning-like** manner. Others are solitary. Some of the largest solitary corals are the mushroom corals (*Fungia* and *Heliofungia*). Due to its nature of extending its feeding tentacles during the day, *Heliofungia actiniformis* is often thought to be a sea anemone and, although it does look like one, it has a stony coral skeleton.

Temperate waters also have beautiful solitary corals with the bright green *Scolymia* and the deeper water dwelling *Balanophyllia* which has fluorescent symbiotic bacteria living around its mouth.

The best time to see corals is during night dives as all have their tentacles out feeding on plankton. The most brilliantly-coloured coral polyps belong to Faulkner's coral (*Tubastrea*) which grows on the roofs and sides of caves and under ledges or jetties in small fist-sized clumps. Sometimes these can be seen out feeding on overcast days, in shade, or on incoming tides.

Faulkner's coral polyps Tubastrea faulkneri (PNG)

CALCIUM CARBONATE SKELETONS

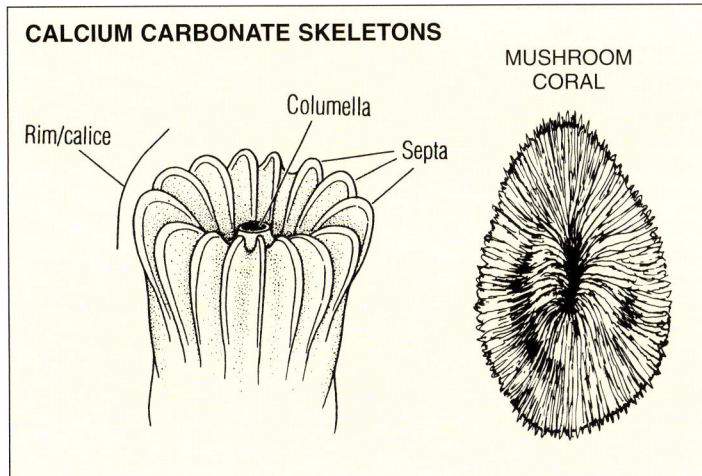

MUSHROOM CORAL

Rim/calice
Columella
Septa

SOME CORAL SKELETONS

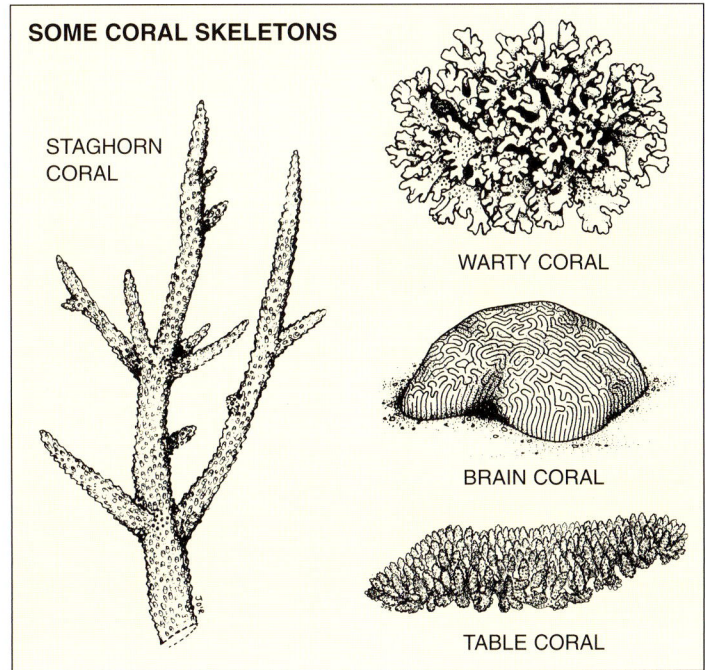

STAGHORN CORAL

WARTY CORAL

BRAIN CORAL

TABLE CORAL

Yellow black coral *Antipathes* sp. (Indo.)

Pipe-cleaner black coral
Antipathes sp. (Mal.)

White black coral *Antipathes* sp. (LHI)

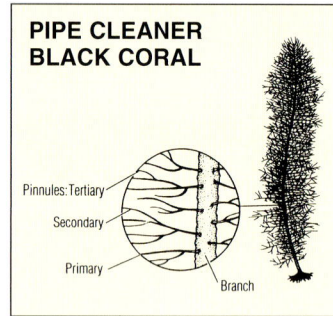
PIPE CLEANER BLACK CORAL

Pinnules: Tertiary
Secondary
Primary
Branch

Sea whip black coral tentacles
Cirripathes sp. (PNG)

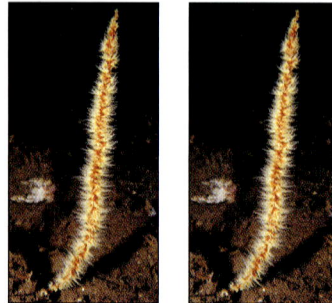
Sea whip black coral *Cirripathes* sp. (Indo.)

Class: Ceriantipatharia (Ceri-anti-path-ar-ia) black corals

Black corals are commonly seen in both shallow and deep tropical Asia/Indo-Pacific waters and in Australia there are at least 20 species. In temperate seas, they are not commonly seen by divers in normal diving situations as they grow beyond the usual recreational diver depth recommendations (40 metres). Also in the temperate seas *Antipathes* appear to be the dominant form.

Living black corals are covered by a **fleshy skin** which may be white, pink, yellow, brown, orange, or green. The **polyps** have **six tentacles** that **do not retract** into the **skeleton**, though they may shrink when disturbed. The **inner skeleton** is **black** and comprised of an extremely tough, horny, tremendously pliable **proteinase material** secreted by the polyps. Most black coral forests have been decimated due to their ancient legend as a "precious" coral which at one time brought high prices in the semi-precious jewellery markets and even today is sold in many tourist shops. In Australia, huge numbers of black coral whips were harvested from the shallow waters of the Great Barrier Reef. When the skeletons of the shallow water living black coral cracked from drying out, our entrepreneurs filled up the cracks with silver solder and sold the polished results as silver-inlaid black coral.

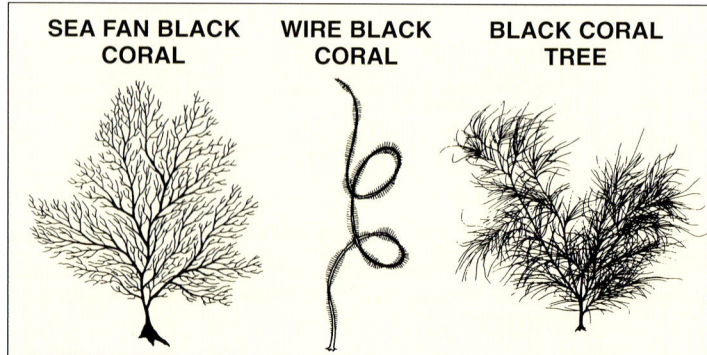
SEA FAN BLACK CORAL   WIRE BLACK CORAL   BLACK CORAL TREE

Class: Ceriantipatharia (Ceri-anti-path-ar-ia) tube anemones

Tube anemones are not well known taxonomically, with new species recently being described from various regions. At present two generic names are in use in the Asia/Indo-Pacific ; *Pachycerianthus* and *Cerianthus*.

Tube anemones are inhabitants of **soft sediments** from low tide level down and beyond 50 metres (130 ft). Each animal constructs a protective, soft mucus tube buried in the mud in which it lives. **Tube anemones** have **two rings of tentacles** around the mouth, with the outer ring tentacles being much longer than the inner tentacles. Although their tentacle nematocysts are able to stun small fish, most of their food consists of plankton caught at night or on dark overcast days, whenever the tentacles are expanded. Tentacle colours may be yellow, pink, brown, white, mauve and grey. They occur in both tropical and temperate waters.

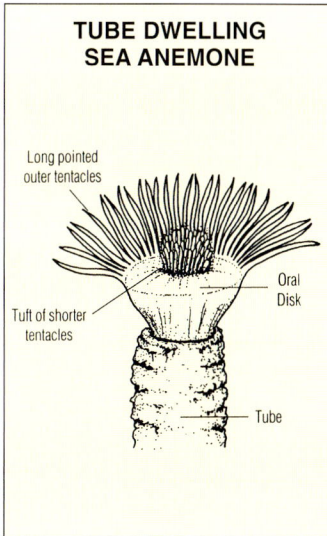

TUBE DWELLING SEA ANEMONE

Long pointed outer tentacles

Tuft of shorter tentacles

Oral Disk

Tube

Tube anemone Pachycerianthus sp.

Phylum: Ctenophora (-ten-o-phor-a) comb jellies

Class: Tentaculata (Ten-tac-u-lata)

Comb jellies were once lumped in with the cnidarians (as Coelenterates)but since it was discovered that they have no stinging cells, science has created a new phylum for them. There are two major types of comb jellies:

(1) The transparent swimming **jelly-like animals** with eight rows of comb-like plates containing fused cilia (which are used for swimming) **live in the water column.**

(2) The second type resembles small flatworms (they can be brightly coloured) and these are generally found living on other invertebrates (sea stars, corals and sea urchin spines). The commensal types are extremely fragile and release long strands of mucus to snare prey. There are thought to be around 100 species, but many more may exist.

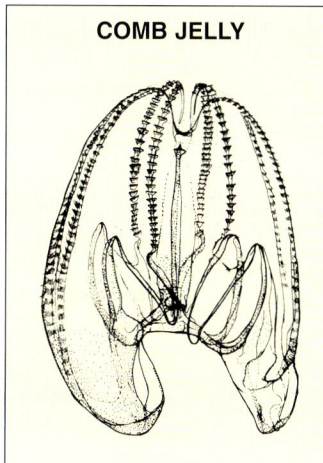

COMB JELLY

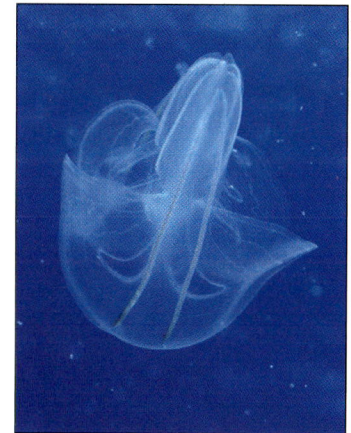

Comb jelly (pelagic variety) (GBR)

Moonscape Ctenophore Coeloplana sp. (PNG)

Sea star Ctenophore Coeloplana sp. (PNG)

Bannworth's Ctenophore Coeloplana bannworthi (Indo.)

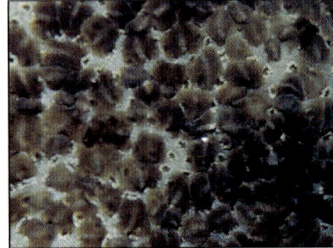
Coral Ctenophore cf Waminoa sp. (PNG)

Phylum: Platyhelminthes (Platy-hel-min-th-es) flatworms

Class: Turbellaria (Tur-bell-ar-ia)

Few divers recognise flatworms and, if they do, they generally see only the more brightly coloured species. Even then they may confuse them with nudibranchs (shell-less molluscs). Although many tropical species may be seen in the open during the day, temperate species are less obvious, living beneath rocks and stones, or hidden among the hollows, or folds of their invertebrate prey. In most cases, a closer look will reveal the difference between a flatworm and a nudibranch. The greater number of marine flatworms are grouped in the order Polycladida. The term "flatworm" includes the tubellarians, or free-living flatworms, parasitic flukes and tapeworms, but is used here to refer to the turbellarians.

Flatworms have **no external gills** (unlike most nudibranchs) but some have **marginal tentacles** at the "head" end which may contain **simple eyes**, and other species may have **dorsal tentacles** issuing from the back near the "head". **Flatworms** are **hermaphrodites**, having complex male and female sex organs. Mating and cross fertilisation occur between two individuals. Eggs are laid on the substrate in spirals, similar to an open-ended circle. Remarkably, the spiral egg ribbons of some flatworms are very similar to those of some nudibranchs.

Most species living in the sea are **predatory** and **carnivorous**. In the author's experience many **flatworms feed** on **colonial ascidians, soft corals** and **bryozoans**. Only a few species show any specific body colour patterns that relate to the markings of the species on which they prey.

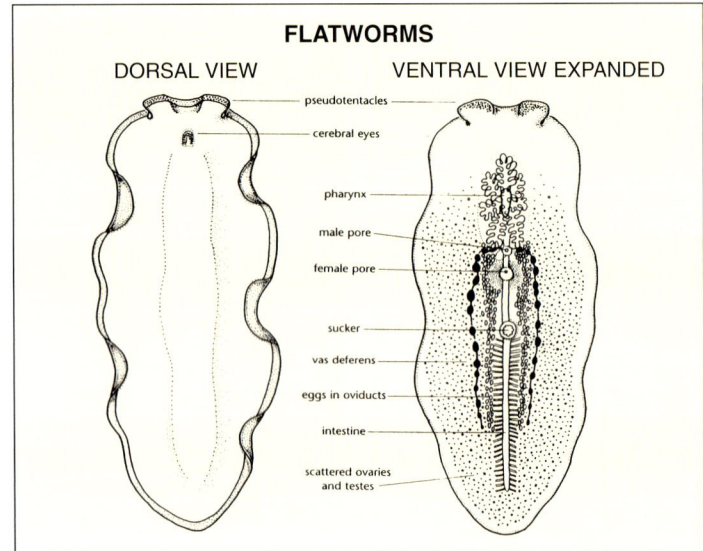

FLATWORMS

DORSAL VIEW | VENTRAL VIEW EXPANDED

- pseudotentacles
- cerebral eyes
- pharynx
- male pore
- female pore
- sucker
- vas deferens
- eggs in oviducts
- intestine
- scattered ovaries and testes

Suzzan's flatworm Pseudoceros suzzanae (Mal.)

Gravier's flatworm Pseudoceros gravieri (PNG)

Linda's flatworm Pseudoceros lindae (PNG)

Phylum: Nemertea (Nem-er-tea) ribbon worms

Class: Anopla

Obscure to most, ribbon worms in general are long, skinny worm-like animals which range in size from several millimetres to several metres in length. They are slippery or sticky to touch and most live beneath rocks or dead coral slabs.

Some are brightly-coloured and some have light sensitive spots (eyes) at the head end. Many are **nocturnal** and only seen on sand at night and even though many of the 1000 species known are indistinct others have colours and patterns which are easy to recognise visually, many species are yet to be described.

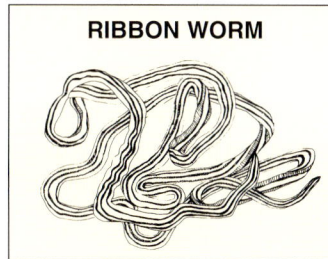
RIBBON WORM

Dash-lined ribbon worm Baseodiscus delineatus (LHI)

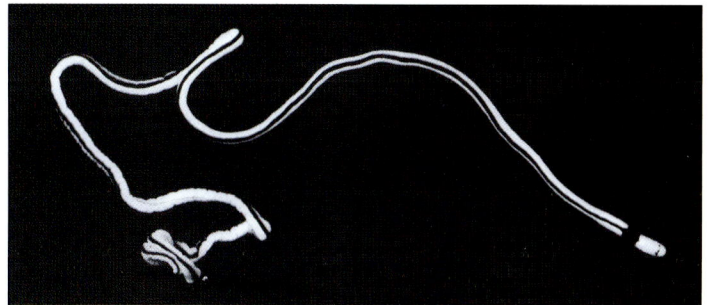
Hemprich's ribbon worm Baseodiscus hemprichii (LHI)

Phylum: Annelida (Ann-el-ida) segmented worms

Class: Hirudinea (leeches)

Although most leeches occur in fresh water there are a number which live in the sea. Most **marine leeches** are known to be **parasitic** and these can be found on many fishes including **sharks, rays** and **flounders**. Most marine leeches have tough cylindrical bodies with a large sucker at the base of the body and a smaller sucker at the tapered head end. Quite often species may be host specific and can be identified this way.

Rayner's shark leech Pontobdella rayneri (GBR)

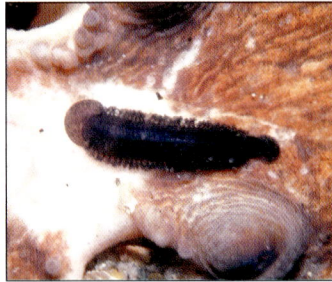

Electric ray leech Branchellion sp. (NSW)

Class: Polychaeta (Poly-kee-ta) polychaete worms

The phylum Annelida, the segmented worms, is one of the major groupings of the animal kingdom. There are more than 12,000 species and 70 families worldwide and comparatively little is known of the natural history or distribution of this diverse group. Ecologically they are of **two types**: the active, **mobile**, foraging Polychaetes, and the **sedentary** burrowing or tube-dwelling suspension or detrital feeders.

Most of the polychaete worms commonly seen underwater are either the larger, free-roving bristle worms, scale worms, and "sea mice", or the stationary deposit-feeding worms, fan worms and tube worms whose brightly coloured feeding tentacles often reveal their presence. Because the head tentacles are the only part of the worm's body that is usually seen, many snorkellers and divers fail to realise that the mystery object is part of a worm, especially when seen in corals (*Spirobranchus*).

Basically, **bristle worms** have **bristles** and **scale worms** have **scales**. **Terebellid worms** inhabit sand or mud tubes on the bottom or beneath rocks and, as deposit feeders, radiate their many retractable buccal (mouth) tentacles over the substrate to collect food. **Serpulid worms** build a hard, calcareous tube, either alone or in a colony. Serpulids are filter feeders, using double branchial crown tentacles and, unlike **Sabellid fan worms**, most have a small, stalked operculum that acts as a trapdoor to seal the tube after the worm retracts.

Reproductive techniques of Polychaete worms are as diverse as their lifestyles and vary from shedding eggs and sperm into the sea to incubating their young in brood chambers. Some carry their developing young on their backs and others carry their eggs only until they hatch, releasing the larvae into the sea, where they grow and develop as members of the **zooplankton** until they are ready to settle and **metamorphose**.

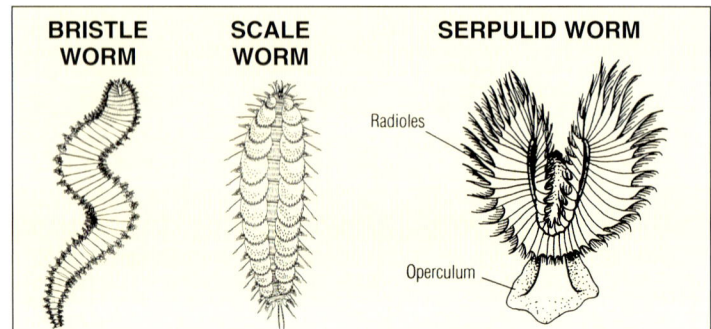

BRISTLE WORM   SCALE WORM   SERPULID WORM — Radioles — Operculum

White-speck bristle worm Euphrosine sp. (PNG)

Tentacular worm Eunice aequabilis (LHI)

Magnificent tube worm Protula magnifica (GBR)

Tangled tube worms Filograna implexa (PNG)

Black-spot scale worm Lepidonotus sp. (PNG)

Club-spined scale worm Gastrolepidia clavigera (Mal.)

FEATHER DUSTER WORM

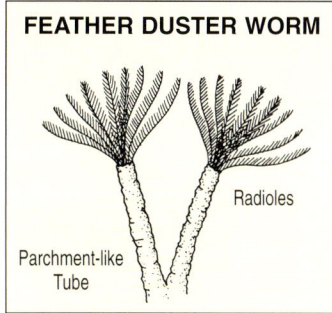

Radioles

Parchment-like Tube

Feather duster worm Sabellastarte sp. (PNG)

CHRISTMAS TREE WORM

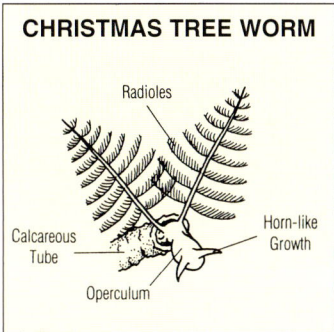

Radioles

Calcareous Tube

Horn-like Growth

Operculum

Christmas tree worms Spirobranchus giganteus (NT)

Spiny-headed tube worm Lygdamis sp. (PNG)

Giant spaghetti worm Reterebella sp. (PNG)

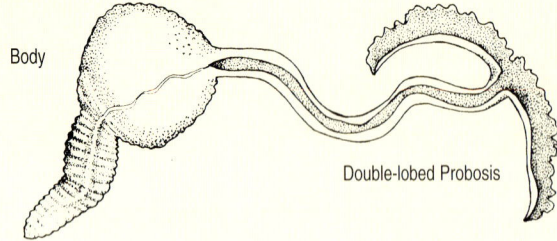

SPOON WORM

Body

Double-lobed Probosis

Green spoon worm Metabonellia haswelli (SA)

Spoon worm tentacles (night) (Fiji)

Evan's sea spider Anoplodactylus evansi (NSW)

Spongy sea spider Nymphopsis sp. (PNG)

Phylum: Echiuroidea (Echi-uroid-ea) spoon worms

Class: (None)

Spoon worms have soft sausage-shaped bodies. They hide under rocks, in crevices under coral or are buried in sediment and extend only a long protrusible proboscis out to collect food particles. The proboscis is generally the only thing observers see and it may be forked or have a flattened, spoon-shaped end. Commonly seen on night dives in the tropics the forked proboscis of a spoon worm has intrigued thousands of divers. As soon as the proboscis is spotlit with a torch beam the creature quickly withdraws. In some species the males are reduced to tadpole-like beings living in the females internal organs.

Phylum: Arthropoda (Arth-ro-po-da) insects

Class: Pycnogonida (Pyc-no-go-ni-da) sea spiders

Similar in shape to terrestrial spiders, pycnogonids generally have eight legs with strong claws at the ends. There are around 1500 described species and none have lungs or an anus; oxygen and wastes diffuse across the cell walls. They occur in most sea bottom habitats and are often found feeding on sea anemones, hydroids or bryozoans which they suck body fluids from. In most species the males carry the eggs and raise the young which ride on their backs.

Long-legged sea spider Nymphon acquidigitatum (Vic)

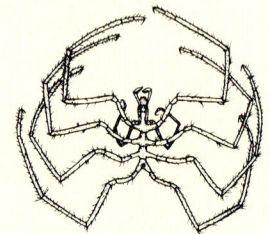

SEA SPIDER

Subphylum: Crustacea (Crust-a-cea) crustaceans
Insects of the sea

Many people are familiar with prawns, crabs and lobsters. Divers often recognise other larger crustaceans like rock lobsters, shovelnose lobsters, blue swimmer crabs and prawns. The smaller crustaceans are less known. There are around 30,500 species throughout the world. **As insects of the sea, their body is encased in an armour-like skin (exo-skeleton). To grow they must shed this skin at various intervals.** During this period they are extremely vulnerable and seek places to hide while their new shell-skin hardens.

It is usually possible to identify crustaceans in the field, though colours and patterns vary, external features on juveniles and adults may differ, and sexual dimorphism is often present. Colour photography is an excellent method of recording crustaceans. They feed on a variety of organisms: bacteria, plankton, sediment, suspended particles, algae, molluscs, fish, worms, other crustaceans and carrion.

The sexes are generally separate and, after mating, the female lays eggs that are carried beneath the abdominal flaps of crabs, on the modified swimmerets (swimming feet) of rock lobsters and shrimps, and on the chests of mantis shrimps. On hatching, the larvae join the plankton and pass through a series of free-swimming stages before settling to the bottom where they metamorphose into juveniles of their particular species.

Crustaceans may be found in almost any habitat in the sea. They live in burrows in sand or mud among rubble, on reefs, in caves and under ledges and rocks. Some bury by day and emerge at night. Some carry shells over their heads and others live in them. Some are permanent swimmers in the vast ocean spaces. There are species only millimetres in size that are carried about in the water column as plankton, and giants weighing 15 kilograms (33 pounds) that crawl about on rocky reefs 100 metres (330 feet) below the surface.

Class: Maxillopoda (Max-ill-o-po-da) barnacles

Very few divers recognise barnacles as crustaceans. After all, who would liken a swift-footed rock crab to a volcano-shaped bunch of shelly plates glued on a rock? Barnacle animals live upside down in their shells and sweep plankton into their mouths with their feet.

Almost all species of barnacles are hermaphrodites though they usually breed by cross-fertilisation, with each barnacle impregnating its neighbour with a penis that can be extended up to 30 times the length of its body. The eggs hatch into planktonic larvae, which (unlike their parents) are immediately recognisable as crustaceans. After a period of feeding in the plankton, a non-feeding larval stage develops that is specially adapted for the critical task of habitat selection. Once settled on a suitable substrate it is there for life.

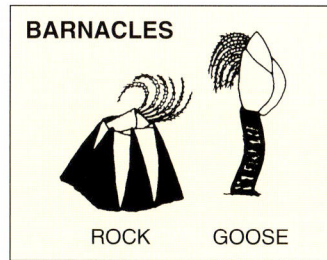

BARNACLES

ROCK   GOOSE

Goose barnacles Lepas anserifera (PNG)

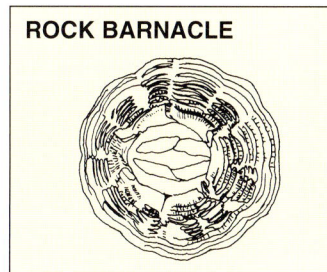

ROCK BARNACLE

Rock Barnacle Catomerus polymerus (Tas.)

Javelin mantis shrimp Alima laevis (NSW)

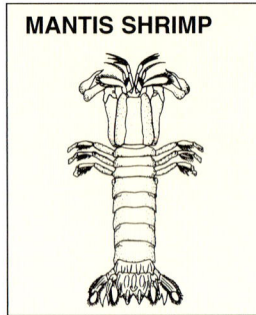
Serrated amphipod Cerodocus serratus (Vic)

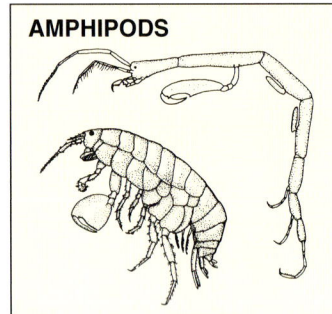
Kapala isopod Bathynomus kapala (NSW)

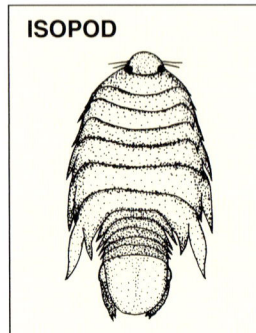
MANTIS SHRIMP

AMPHIPODS

ISOPOD

Class: Malacostraca (Mal-a-cost-ra-ca) mantis shrimps, amphipods, isopods, shrimps, rock lobsters, hermit crabs, squat lobsters, crabs

Stomatopods (Stom-at-o-pods) mantis shrimps

Superficially similar in appearance to large prawns, **mantis shrimps** have a **long flat body** with a short carapace and eight appendages, six legs and variously modified **raptorial claws**. These claws are reminiscent of **preying mantis** of the **insect world** and each species of **mantis shrimp** is armed with specific versions depending on their prey. There are fish stabbing and grasping versions which have sharp points and rows of needle-sharp spines. Those that feed on bivalve and univalve shells have blunt swollen claws with bevelled edges.

Amphipods (Am-fi-pods)

In general, amphipods have small, flattened, laterally compressed bodies often protected by large plates or scales along the sides. Eyes are set on the sides of the head with large or small antennae and no carapace. The first and second pair of thoracic legs have pincers or claws for feeding and grasping. The last three pairs of backward angled legs are generally used for walking, jumping or pushing. There are around 8000 described species known across the world. Amphipods exploit every major aquatic habitat.

Isopods (Iso-pods)

The typical terrestrial slater or fish louse isopod has a dorsally compressed body generally protected by well-developed scale-like plates. Devoid of a carapace the first pair of antennae are smaller than the second pair and the unstalked eyes are on either side of the head. The first pair of grasping thoracic limbs often used as secondary mouth parts. Within this order there is a great deal of body form diversity. Most isopods are free-ranging, bottom-dwelling creatures occupying a bewildering number of habitats, from intertidal areas down to 5,000 metres. Some live out their lives as entrenched parasites burrowed into the bodies of fish.

Shrimps and prawns

Prawns and shrimps are found throughout the world with most of the 1000 or more species living in tropical seas. In general, the **rostrum** is well developed, usually **serrated** and the feelers are similar in size; **nocturnal species** have **larger eyes.**

Prawns are strictly marine, live mostly on soft bottom (sand or mud) and are mostly seen at night. Shrimps may be freshwater, or marine where they live in a variety of habitats including coral reefs, rocky reefs, rubble, seagrass meadows and down burrows in sand, under rocks in caves, under ledges. With such a variety of lifestyles food ranges from algae, to detritus and may include bacteria, plankton, mucous, copepods, fish tissue, sea stars and scavenged material.

Colour patterns can be variable between males and females (sexual dichromatism) but as many pairs are found together identification is not difficult on known forms. Many species form commensal relationships with other sedentary animals (sea anemones).

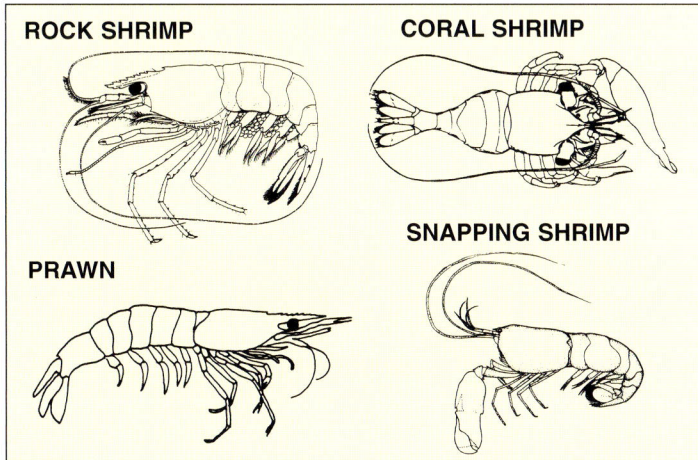
Common snapping shrimp
Alpheus strenuus (NSW)

Harlequin shrimp Hymenocera picta (PNG)

School prawn Penaneus sp. (night) (PNG)

ROCK SHRIMP

CORAL SHRIMP

SNAPPING SHRIMP

PRAWN

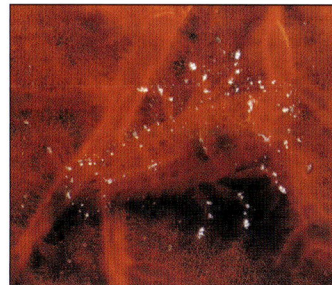
Paron's sponge shrimp
Gelastocaris paronae (PNG)

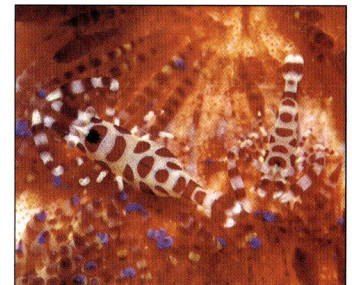
Coleman's shrimp Periclimenes colemani
(Phil) (Photo: Roger Steene)

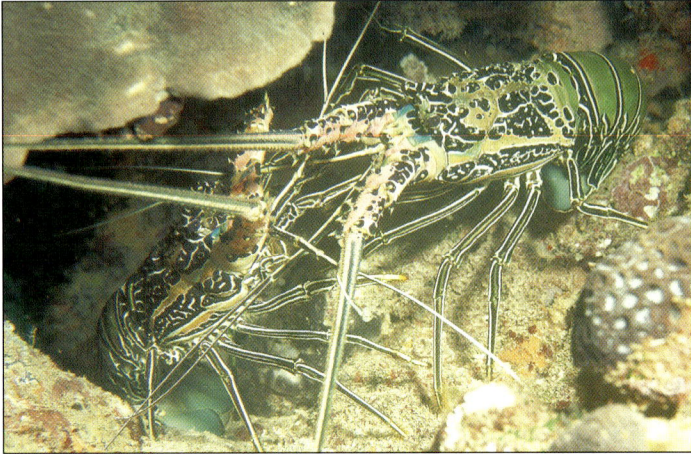
Painted rock lobsters courting Panulirus versicolor (WA)

Slipper lobster Parribacus antarticus (CS)

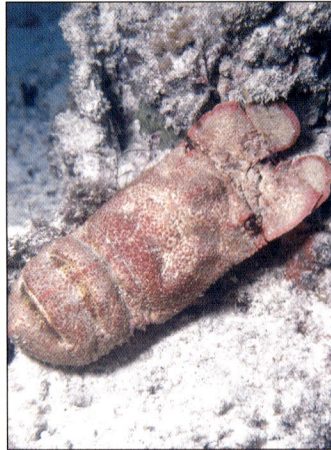
Squamose slipper lobster Scyllarides squammosus (GBR)

Rock lobsters and slipper lobsters

Spiny crayfish or rock lobsters are well known throughout most tropical seas where they form the basis of major fisheries. As adults these animals are fairly large (400 mm) with a **cylindrical-shaped carapace**, a pair of **long spiny antennae** and eyes which are high on the head and protected by **large curved spines** (rostral horns). The powerfully muscled abdomen has downward pointing curved spiny plates along the edges and the tail is broad and strong.

Rock lobsters usually live in caves, under large coral heads or table corals. Nocturnal of habit they venture forth during the night to hunt prey (molluscs).

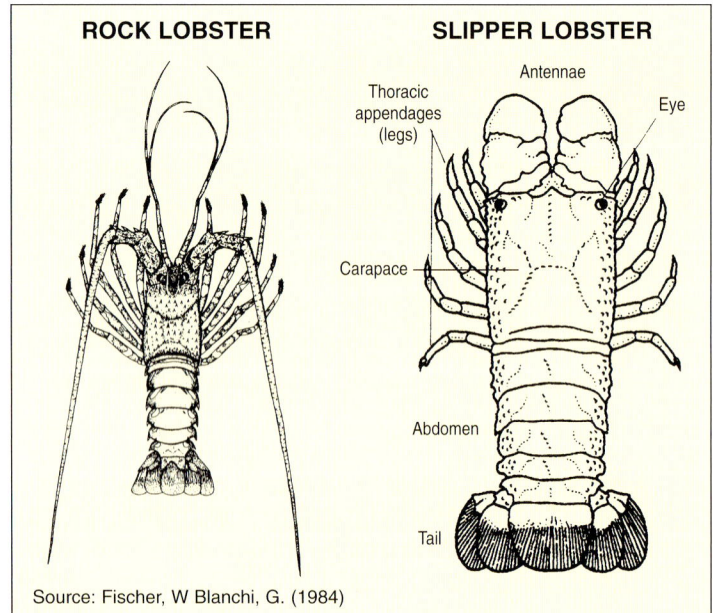

ROCK LOBSTER   SLIPPER LOBSTER

Antennae

Thoracic appendages (legs)

Eye

Carapace

Abdomen

Tail

Source: Fischer, W Blanchi, G. (1984)

Hermit crabs

Most **hermit crabs** have **no protective shell** for their abdominal soft parts, they utilise the shells of dead **univalves** in order to survive and as such are **not true crabs**. The giant land hermit crab known as the coconut crab (*Birgus latro*) only uses univalve shells at a juvenile stage developing a hardened "shell" for its soft parts as it grows. Most **hermit crabs** have short strong legs, **long eye stalks** with **small eyes** for **diurnal kinds** and **larger, round eyes** in **nocturnal species**. The **diogenids** and **coenobitids have larger left hand claws** which they use to plug the aperture of the mollusc shell and **pagurids have a larger right hand claw**.

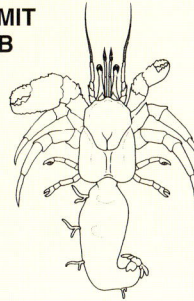

Hairy hermit crab Dardanus lagopodes (LHI)

Anemone hermit crab
Dardanus pedunculatus (nocturnal) (LHI)

Siagian's squat lobster Lauriea siagiani
lives on sponges (Indo.)

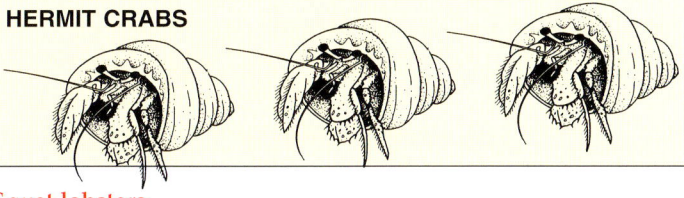

Squat lobsters

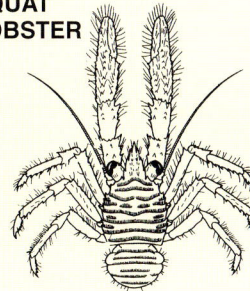

Small lobster-like crustaceans belonging to the family Galatheidae occur throughout the Asia/Indo-Pacific but as they are quite small (15 to 20 mm) few people notice them. They have a **flattened carapace**, a **well-developed rostrum** and a **symmetrical abdomen** which is **curled beneath the body**. Their large **claws** are **elongated** and the **last pair of legs** are generally **under-developed stubs**.

Species of the genus *Munida* have **large eyes** and a **triple-spined rostrum** in which the **central spine** is **larger** than those on each side.

Almost all species have distinctive shapes and colour patterns though few have been published in colour and a lot more remain undiscovered. They can be visually identified from a photograph but initial identity must first be confirmed by a taxonomist from a specimen.

Elegant squat lobster Allogalathea elegans
lives on feather stars (Indo.)

63

Coleman's porcellanid Petrolisthes sp. (Indo.)

Rough spider crab Schizophrys aspera (LHI)

Red-back swimming crab Portunus sp. (PNG)

Beautiful reef crab Lophozozymus pulchellus (WA)

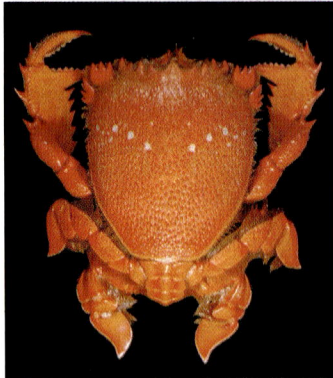
Spanner crab Ranina ranina (Qld)

Toothed sponge crab Cryptodromia octodentata (SA)

Crabs

Over 6000 different kinds of crabs occur throughout world oceans. Crabs can walk, run, crawl, clamber, swim, dig, burrow and scurry depending on their type. Although some species have evolved into incredibly bizarre examples they all retain the same basic body form. The **body** is encased in a **carapace** with a **reduced abdomen** folded under as an **abdominal flap** housing the sexual organs. Most have **claws** known as **chelae**, these may be exaggerated in size and are used for **defence**, to capture prey, attract females and repulse other males.

Due to their shapes and patterns the identification of some crab types is fairly simple and many can be determined to species once their colour images are available. The commensal species (wherever the association is well known) are also fairly easy to identify; others require the lengthy process of tracing the specimen or picture by process of elimination in scientific papers.

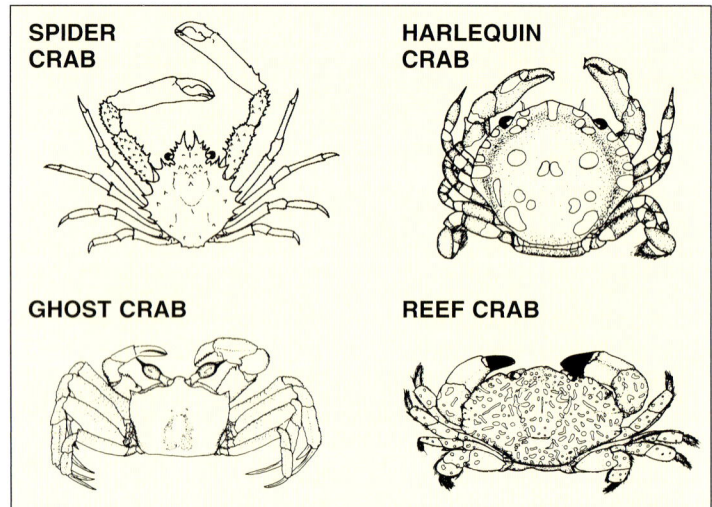
SPIDER CRAB

HARLEQUIN CRAB

GHOST CRAB

REEF CRAB

Phylum: Mollusca (Moll-us-ca) molluscs - shells

The molluscs make up a major part of the world's marine invertebrate faunas. At least 112,000 species of molluscs are known and probably many more remain to be discovered. The word mollusc means soft-bodied which is applied to all five major classes within the phylum.

Class: Polyplacophora (Poly-plac-oph-ora) chitons

Chitons are sometimes called "coat-of-mails" shells due to their **eight-valved shells** set in a tough "leather-like" girdle being similar to armour worn by ancient soldiers. **Chitons** can also roll up into a ball when disturbed, a bit like the defence mechanism displayed by the South American armadillo. **Chitons** are all **marine** and most species are intertidal. Subtidal forms live beneath rocks and dead coral. They are slow moving and are able to clamp their bodies to the substrate so tightly that it is difficult to remove them without damaging the animal.

Owing to their cryptic habits many chitons are not well known. *2002 Sea Shells* (**Catalogue of Indo-Pacific Mollusca**), Neville Coleman, has the first major coverage of Indo-Pacific chitons in colour.

Class: Gastropoda (Gas-tro-po-da) univalves, snails

The class **Gastropoda** contains **more species** than any other within the phylum Mollusca and up to 90,000 species are thought to exist worldwide. Species are fairly well known and numerous colour guides have been published since the 18th century. This diverse class requires listing to subclass level to simplify description.

Subclass: Prosobranchia (Pros-o-branch-ia) shells snail-like, generally coiled: cones, cowries, helmets, tritons, murex, winkles, whelks, top shells, etc.

The molluscs of this subclass have gills and anus in front of their visceral mass, which is sometimes referred to as the visceral hump.

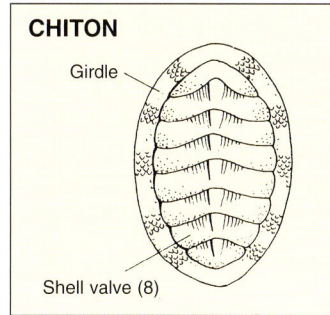

CHITON

Girdle

Shell valve (8)

Oak Chiton Onithochiton quercinus (Qld)

GIANT TRITON (univalve)

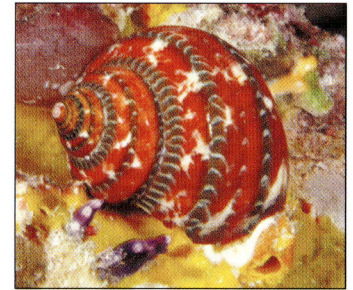

Cat's-eye turban Turbo petholatus (PNG)

Sieve cowry Cypaea cribaria (Van.)

Oriental moon snail Naticarius orientalis (PNG)

Wavy-lined bubble Micromelo undatus (Qld)

BUBBLE SHELL

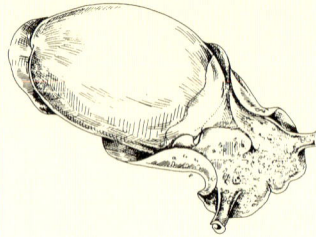

Juliana sea hare Aplysia juliana (LHI)

SEA HARE

Forskal's pleurobranch Pleurobranchus forskalii (Nor Is)

PLEUROBRANCH
Side-gilled slug

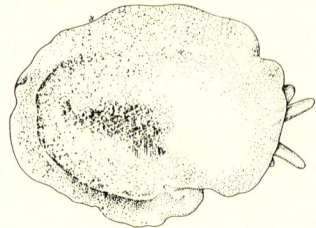

In general they have a **single shell coiled into a tube, similar to a snail.** The tube is open at one end and may be partly sealed by a trapdoor (operculum). The head is well formed, with two eyes, each set at the base of a tentacle and may have a **retractable proboscis**, or "trunk", containing a well-developed **radula** (multiple teeth on a flexible ribbon). Most of the larger species have characteristic features and many thousands can be visually identified.

Subclass: Opisthobranchia (O-pisth-o-branch-ia) shell reduced, sometimes internal, or absent; nudibranchs, pleurobranchs, sea butterflies, bubble shells, sea hares, etc.

Entirely marine, this subclass offers the greatest challenge to the marine naturalist. The opisthobranchs are the "butterflies" of the sea and are among the most beautiful and spectacular underwater creatures. Over 3000 species are known across the world.

These molluscs may have reduced external shells, internal shells, or no shells at all in their adult form. They may have gills behind or beside their visceral mass, or they may have no gills at all, depending on the respiratory functions of the skin, or on the other parts of the body (such as pustules or cerata). Many have a well-defined head, one pair of eyes and at least one pair of tentacles. The eyes are reduced, or non-existent in some species, with most only acting as light receptors.

They are generally active, mobile animals, and some are even able to swim. With specific plant or animal diets many are seasonal in their appearance and live in almost every habitat available in the sea. Some are easy to see and advertise their presence with blatantly brilliant colours hiding their repugnant and bad tasting properties. The bright colours and patterns serve as a warning to visual predators such as fish that learn not to mouth these animals. Most species (where known) can be identified from a good transparency, digital image or colour reproduction (see *1001 Nudibranchs* **(Catalogue of Indo-Pacific Sea Slugs)**, Neville Coleman, for 1700 pictures).

Spanish dancer Hexabranchus sanguineus (WA)

Class: Scaphopoda (Scaph-o-po-da) tusk shells

All the inhabitants of this class live in sand or mud and are usually only seen in beach drift. The shells are tusk-like and their animals feed on microscopic life forms in the sand. The tusk shell mollusc resides head down in the sand or mud at an oblique angle, with the posterior end of the shell projecting from the sand. Tusk shell animals have no eyes, no gills and no heart. Their nervous system is rudimentary, as is their circulatory system, but strangely enough they can "hear" or detect vibrations. Tusk shell animals are either male or female and eggs and sperm are released into the water, where fertilisation occurs.

Species identification is sometimes difficult due to lack of easily seen characteristics in smooth forms, others have very defined shapes, colours and patterns.

Glenie's Chromodoris Chromodoris gleniei (Mal.)

Baba's Phyllidia Phyllidia babai (PNG)

Orange-backed Doto Doto sp. (LHI)

Much-desired Flabellina Flabellina exoptata (Qld)

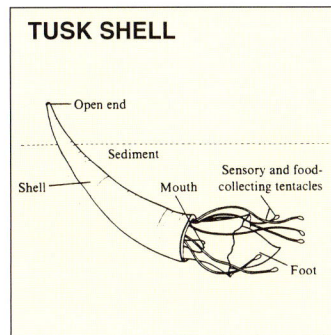

TUSK SHELL

Open end

Sediment

Shell

Mouth

Sensory and food-collecting tentacles

Foot

Beautiful tusk shell Pictodentalium formosum (Jap)

Ballot's saucer scallop Amusium balloti (Qld)

Lamellate venus shell Antigona lamellaris (NSW)

Reef-flat octopus Octopus sp. (WA)

Little reef squid Sepioteuthis sp. (PNG)

Pearly Nautilus Nautilus pompilius (Van)

Paper Nautilus Argonauta nodosa (Tas)

Class: Bivalvia (Bi-val-via) bivalves

Bivalves can be easily recognised because they almost always have **two shell valves**. Each valve is hinged at the dorsal margin; the animal inside is laterally compressed and can open or close the shell valves by contracting or relaxing the adductor muscles attached to the inside of each valve. Many bivalves are attached or fixed to the substrate, either permanently like oysters, or temporarily like some mussels; others (cockles) are free living.

Some scallops can swim and are propelled by water jetted out of the mantle cavity but this mobility is rare. Most burrowers can move along and dig by means of a well-developed foot. Bivalves with colour patterns, designs, or other distinguishing features such as shape, spines, ridges, bumps, or nodules are fairly easy to recognise to species level.

Class: Cephalopoda (Keph-al-o-po-da) (means "head-footed") cuttlefish, squid, octopus, nautilus, argonauts

Apart from the ram's horn squid, nautilus, and argonauts (egg case), the cephalopods in general do not produce shells. Octopods have no shell at all and almost all the squids have but a thin rudimentary internal "pen" or gladius. In the cuttles this "pen" is a much heavier structure and cuttle bones from dead animals are one of the most familiar objects in the flotsam cast ashore by the waves.

Cephalopod colouration is regulated by pigment cells (chromatophores) in the skin. These cells can be expanded or contracted by the cephalopod at will. By rapid reduction and expansion of the pigment cells the animal can cause instant colour changes to take place. Although stimulation of substrate colouration and texture can be assumed by cephalopods, squids, cuttles and octopus seem to use definite inherent patterns. Each pattern has a particular purpose, and these include warning patterns, sex patterns, mating patterns, hunting patterns, and escape patterns.

Cuttle bone Sepia sp. (NSW)

Port Hughes cuttlefish Sepia sp. (SA)

Wide-banded cuttlefish Sepia latimanus (Van)

Phylum: Brachiopoda (Brach-io-po-da) brachiopods, "lamp shells"

Class: Articulata

With over 30,000 species known from fossil records brachiopods were once very common. there are only thought to be around 350 species alive in world oceans today. Although brachiopods appear superficially similar to bivalve mollusc shells their anatomy is quite different. They live in the dark attached beneath caves and ledges and under rocks by a muscular stalk (peduncle). Brachiopods are filter feeders and reproduce by males and females releasing reproductive products into the surrounding water.

Phylum: Phoronida (Phor-o-ni-da) phoronids, horseshoe worms

Class: (None)

Wormlike animals that live in parchment tubes in soft bottom areas of the sea floor, there are only around 15 species worldwide Superficially similar to fan worms, they are generally black, grey or white in colour and their ciliated tentacles are arranged in a horseshoe shape around the mouth. Indo-Pacific species are often found inhabiting the soft tubes of burrowing tube anemones in a commensal relationship (*Cerianthus*). They reproduce by shedding sperm and eggs into the water column. The larvae drifts with the plankton for some time before settling in a suitable habitat.

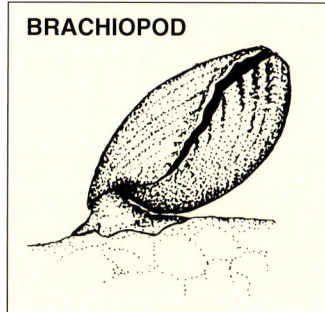

BRACHIOPOD

Red-dappled brachiopod Frenulina sanguinolenta (PNG)

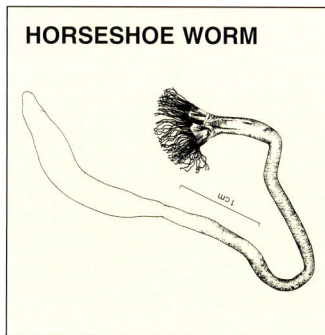

HORSESHOE WORM

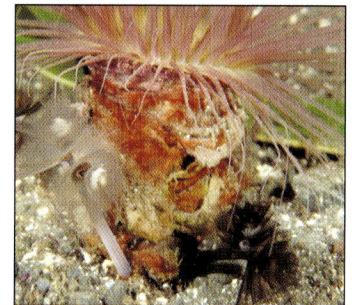

Horseshoe worms Phoronis sp. (PNG)

Flower-petal bryozoan (PNG)

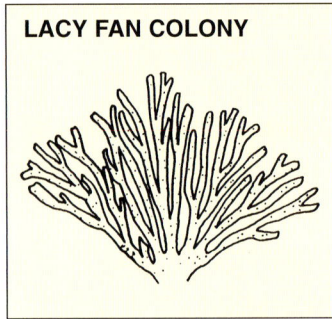
LACY FAN COLONY

Vase-shaped zooid chambered species (PNG)

Sea weed bryozoan Costaticella sp. (LHI)

Purple bryozoan Iodictyum phoeniceum (LHI)

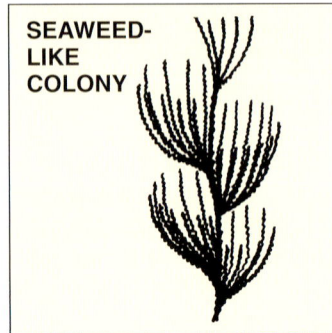
SEAWEED-LIKE COLONY

Phylum: Bryozoa (Bry-o-zo-a) sea mosses

Class: Gymnolaemata

There are over **4000** species of living **bryozoans**, most are marine but fresh water species do exist. All bryozoans and polyzoans (pliable bryozoans) are colonial and grow in an amazing array of shapes, configurations and designs, from small **circular encrusting formations**, intricate **lace-like** structures and **plant-like tufts**, to large clumps of coral-like colonies several metres in circumference. These structures have been known in the past as "lace corals".

The term "sea mosses" relates to the fact that most bryozoan colonies encrust and grow over the surfaces of rocks or other organisms such as algae or shells just as moss may grow on wood or rocks.

Species identification is mostly determined by a trained taxonomist with the aid of a microscope and specimen though some genera are visually recognisable.

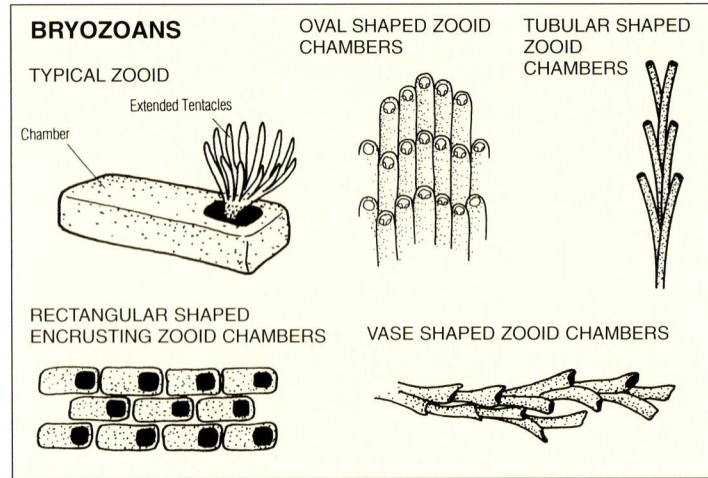
BRYOZOANS
TYPICAL ZOOID — Chamber, Extended Tentacles
OVAL SHAPED ZOOID CHAMBERS
TUBULAR SHAPED ZOOID CHAMBERS
RECTANGULAR SHAPED ENCRUSTING ZOOID CHAMBERS
VASE SHAPED ZOOID CHAMBERS

Phylum: Echinodermata (Echin-o-derm-ata) spiny skins

Echinoderms

The term echinoderm means "spiny skin" and in this phylum there are **five major classes**: sea stars, brittle stars, feather stars, sea urchins and sea cucumbers. Rarely would a sea dive pass without a diver encountering some form of echinoderm on the bottom. Many species are brightly coloured, moderately sized, simple to recognise and some have sharp spines (sea urchins).

Although this group is extremely varied in their structure, there is one feature by which they are all united. This character is called penta-radial symmetry. The **adult body** is **divided** into **separate but similar sections located around a central process.** Most have **five sections** and so being are called **pentamorus** (meaning five) a design **unique** in the animal kingdom.

Class: Crinoidea (Crin-oid-ea) feather stars

Very common in ancient seas (over 400 million years ago) the stalked sea lilies are still present in modern oceans but are restricted to deep waters, or in the cold depths beneath the ice in Antarctica. Whereas the **stalked sea lilies** are **anchored** in the **bottom sediments** and do not move around, the **feather stars** can crawl, roll, walk and even swim short distances when **moving** to better feeding situations.

The body of a typical **feather star** is cup-shaped, supported on a circular arrangement of jointed appendages known as **cirrus** which **cling** to the **substrate** and are attached to a base on the underside of the cup. The sides of the cup extend up into **five many-jointed arms** which then (depending on the species) fork at various intervals and **produce multiple branches**, or arms.

Although many **feather stars** can be seen with their **arms extended** and **feeding** during the day, at night many more come out of the reef to move around and feed. On some reefs there are extensive populations while others have less, depending on the current

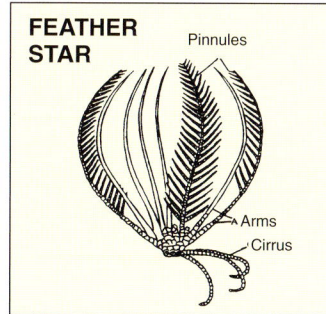

FEATHER STAR
Pinnules
Arms
Cirrus

Stalked juvenile feather stars, still attached to their settlement substrate

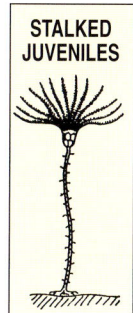

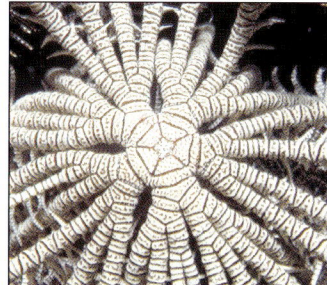

STALKED JUVENILES

Underside of Feather star without cirrus

Feather stars without cirrus, cling with their arms (Van.)

Feather star with well-developed cirrus

Robust feather star Himerometra robustipinna (WA)

Spiny sand star Astropecten vappa (WA)

SEA STAR

Five Arms

Arms merge
at base into
Central Disk

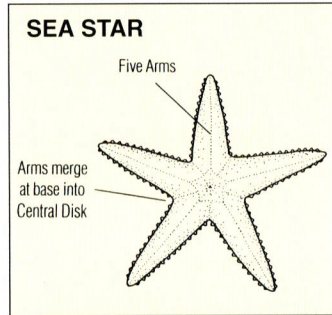

Nodose sea star Fromia nodosa (Mal.)

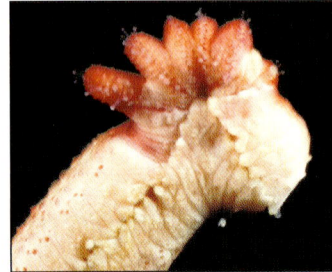
*Asexual reproduction by autonomy
(cast off arm) Echinaster luzonicus (PNG)*

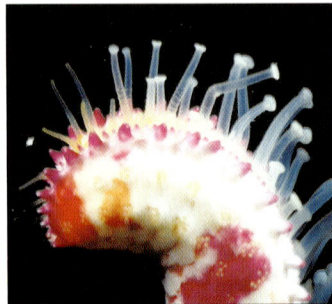
Tube feet (surface dwelling sea stars) (NSW)

Tube feet (burying sea stars) (NSW)

movement. Feather stars do not appear to have many predators. If **damaged** they can replace their **arms at will.**

Most feather stars require a trained taxonomist for identification. However, with more information becoming available together with increased photographic coverage this should become easier in the future.

Class: Asteroidea (As-ter-oid-ea) sea stars

As these **animals** are **not fish**, it seems rather ridiculous to continue using the outmoded name "starfish" from the annals of ancient history. **Sea stars** are very common, from the **intertidal zone** down to and beyond 100 metres, they **inhabit** almost **every major habitat** from coral reef to muddy bottom. While most species have the **typical five arm design**, others may have up to 13 arms (crown of thorns). Even normal five-sided sea stars can be found as three-armed, four-armed or six-armed on occasions.

All echinoderms have tube feet modified to their way of life. **Reef** and **sand surface-dwelling** sea stars have **tube feet tipped with suckers** which enable them to hold onto reefs during rough seas, transverse over any terrain (even upside down) at a fair pace and catch and hang onto prey. In the case of bivalve prey the suckered arms are used to bring pressure to bear on each side of the bivalve causing it to open enough to allow the sea star's extruded stomach to dissolve and digest the bivalve animal.

Sub-sand dwelling sea stars (*Astropecten*) have **tube feet** that are **pointed**, especially modified for digging. When disturbed or threatened on the surface of the sand, these sea stars can virtually 'melt' below the surface right before the diver's eyes. Sub-sand dwelling sea stars swallow small prey whole and may have as many as five bivalves in their mouth at once.

Sea stars as a group are **carnivorous, herbivorous, omnivorous** and also **scavenge**; some are **cannibalistic**. Most species can be identified from a good colour reproduction.

Class: Ophiuroidea (Ophi-ur-oid-ea) brittle stars, snake stars, serpent stars, basket stars

A **basic brittle star** is comprised of a **central disc-like body** from which **five snake-like arms** protrude. The body disc generally appears quite soft in brittle stars and serpent stars, but may be heavily armoured, with well developed plates in basket stars. The arms have no obvious ambulacral grooves on the underside and may be plated, ridged, knobbed, spined or grooved depending on the species. These arms and their particular arrangements are most important to the brittle star as they perform both locomotion (brittle stars are the most active and fastest moving echinoderms) and feeding functions. The **tube feet** situated along the arms **play no part in brittle star mobility**, but instead are **primarily used for feeding**.

With enough references on habitat, colour, patterns and close-up photography it is possible to identify quite a number of brittle stars. However, many species are extremely variable in colour and pattern and juvenile forms may appear different to adults. Most require a specialist taxonomist to identify them from a preserved specimen.

BASKET STAR

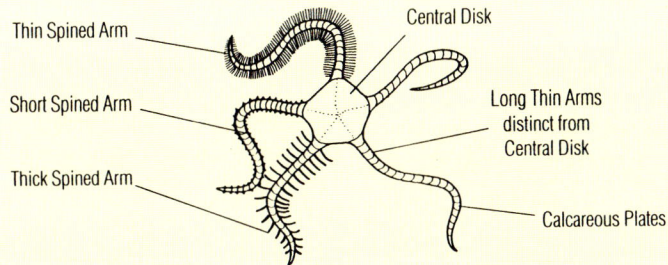

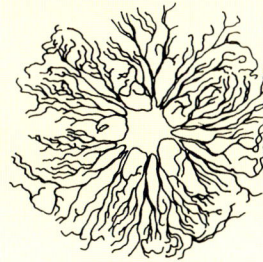

Ludwig's basket star Conocladus ludwigi (Indo.)

Grainy serpent star Cryptopelta granifera (PNG)

Snake star Astrobranchion constrictum (NSW)

BRITTLE STAR

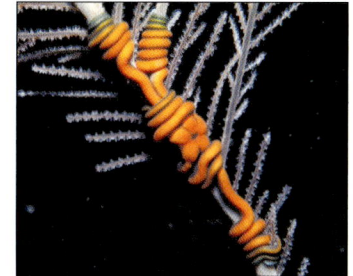

Thin Spined Arm
Central Disk
Short Spined Arm
Long Thin Arms distinct from Central Disk
Thick Spined Arm
Calcareous Plates

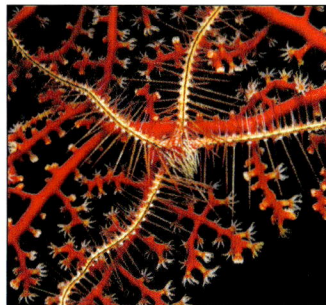

Sea fan brittle star Ophiothrix sp. (GBR)

Purple-banded brittle star Ophiothrix nereidina (PNG)

Savigny's sea urchin Diadema savignyi (Qld)

Fire urchin Asthenosoma varium (Indo.)

Globular urchin Mespilia globulus (PNG)

Slate pencil urchin
Heterocentrotus mammillatus (PNG)

Southern heart urchin Breynia australasiae
(LHI)

Lesueur's sand dollar Peronella lesueuri
(Jap)

Class: Echinoidea (Ech-in-oid-ea) sea urchins

Perhaps for more than any other outstanding feature, **sea urchins** are mostly **recognised** as such by their **spines**, which in many species are **sharp**, pointed and, in some cases, **venomous**. The **skeleton** of a sea urchin is made up of small plates of **calcium carbonate** fused together to form a body shell or test covered by thin skin. Similar to most echinoderms **regular sea urchins** have **radial symmetry**. The test may be circular in shape (like an inflated ball with the top and bottom flattened) or in the case of **irregular sea urchins** such as **heart urchins** or **sand dollars** (which live beneath the sand) the **shape** is **modified** for movement within this habitat.

Most tropical sea urchins can now be identified from a good colour slide. However, some require identification by a trained taxonomist.

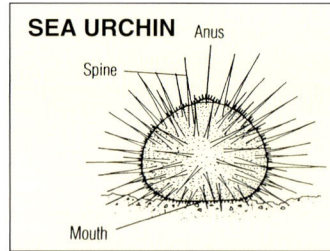
SEA URCHIN
Anus
Spine
Mouth

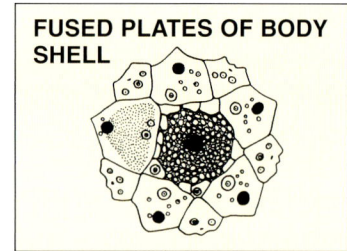
FUSED PLATES OF BODY SHELL

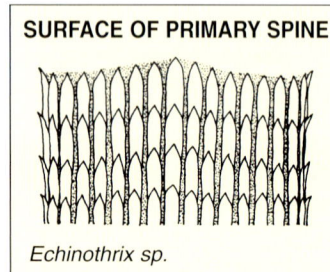
SURFACE OF PRIMARY SPINE
Echinothrix sp.

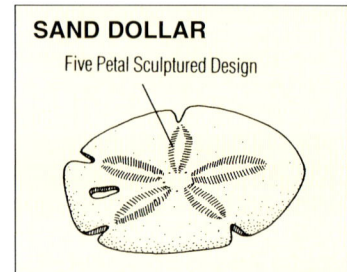
SAND DOLLAR
Five Petal Sculptured Design

Class: Holothuroidea (Holo-thur-oid-ea) sea cucumbers

The majority of **holothurians** are shaped like **cucumbers** or **sausages** and, although some (when alive) feel firm to the touch, their bodies are comprised of **soft tissue** which is somewhat supported by the presence of **minute spicules** of **calcium carbonate** embedded in the body walls. These spicules are used by taxonomists to separate one species from another, as each species has a suite of characteristic patterns. Defensive Cuvierian organs are ejected from the anus of some species.

Sea cucumbers may have tube feet either covering their entire bodies, or in rows along their underneath side to aid in locomotion or attachment, or have entirely naked or smooth bodies. While other echinoderms have a **radial symmetry**, holothurians have **bilateral symmetry** with a distinct dorsal and ventral side.

Traditional identification is by microscopic study of the various shaped spicules in each sea cucumber's body walls. However, once a good colour photograph (or series) showing the animal in its habitat is available for comparison, field identification for many species is achievable to an experienced eye.

Pineapple sea cucumber Thelenota ananas (PNG)

Sticky Cuvierian organs Bohadschia marmoratus (PNG)

Red-lined sea cucumber Thelenota rubralineata (PNG)

SEA CUCUMBERS

Mouth

Podia

Anus

Tubed feet with suction disks

MICROSCOPIC SPICULES AND RODS FROM BODY WALLS OF SEA CUCUMBERS

Stars and stripes sea cucumber Pseudocolochirus violaceus (WA)

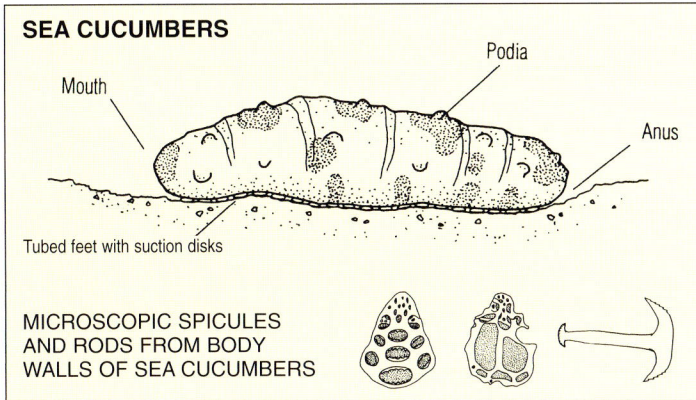
Lampert's sea cucumber Synaptula lamperti (PNG)

75

Phylum: Chordata (Chor-da-ta) animals with backbones

Class: Ascidiacea (Ascid-ia-ce-a) ascidians, tunicates, sea squirts

There are at least **2000** types of **ascidians** in the world's ocean. They are entirely marine and live at all depths from intertidal down to the abyssal trenches. They can be found on reefs, rubble, algae, seagrasses, shells, rocks, in caves, under ledges and in sand and mud. They live as **solitary** or **colonial** animals **attached** to the **substrate** for their entire adult life. A colony grows as the small individuals in it replicate (by budding or subdivision).

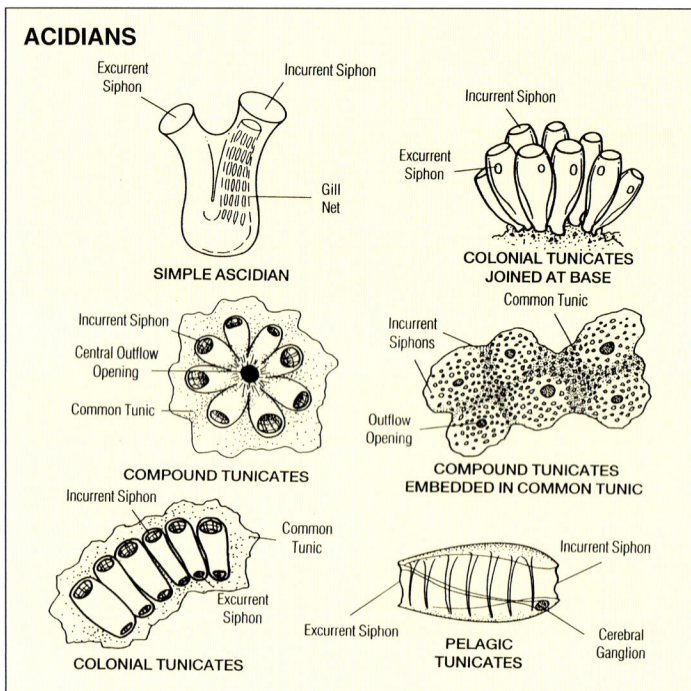

ACIDIANS

SIMPLE ASCIDIAN
- Excurrent Siphon
- Incurrent Siphon
- Gill Net

COLONIAL TUNICATES JOINED AT BASE
- Incurrent Siphon
- Excurrent Siphon

COMPOUND TUNICATES
- Incurrent Siphon
- Central Outflow Opening
- Common Tunic

COMPOUND TUNICATES EMBEDDED IN COMMON TUNIC
- Common Tunic
- Incurrent Siphons
- Outflow Opening

COLONIAL TUNICATES
- Incurrent Siphon
- Common Tunic
- Excurrent Siphon

PELAGIC TUNICATES
- Incurrent Siphon
- Excurrent Siphon
- Cerebral Ganglion

Solitary ascidians just increase in size and shed their outer skin (tunic) on a regular basis.

Sea squirts, or ascidians, may not be the most well known underwater creatures yet they are the **most advanced** of **invertebrates**. Their singular unique character is a **perforated pharynx**, a **sieve** that traps their food as they pump water in through the mouth. Their other **body opening** is an aperture **ejecting** spent water **waste products**, eggs, sperm or larvae. The larvae are tadpole-like and have a notochord, or rudimentary backbone.

Sea squirts (including tunicates) are enclosed in an external coat (or tunic) of a cellulose-like substance (called tunicin) unique in the animal kingdom. **Solitary animals** are usually large, with a sac-like body with their gut and gonads tucked up beside the large perforated pharynx and the mouth opening is larger than the exhalant opening. Many species live **communally**, either as **singular tests attached** to each other by a common base (colonial ascidians) or as a colony of separate zooids formed into a firm jelly-like matrix or investing sheet of tunicin (compound ascidians).

Many compound ascidians have exquisite patterns and colours with pigment particles or blood vessels scattered throughout. The small or intake siphons are often formed around a larger communal exhalant siphon (looking like minute rosettes). Although some compound ascidians are superficially similar (visually) to sponges they are easily distinguished from one another.

Ascidian siphons close at any disturbance. Sponges (having no nervous system) show no obvious reaction to disturbance and their oscula openings remain open.

Visual recognition to species level has come a long way over the past 30 years and many ascidians can now be recognised from a good photograph. However, specific identification often depends on dissection of taxonomically preserved specimens and specialised knowledge.

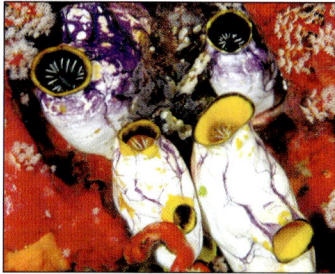
Gold-mouth ascidian Polycarpa aurata (Indo.)

Moluccen ascidian Clavelina moluccensis (PNG)

Coleman's ascidian Polyandrocarpa colemani (Qld)

Magnificent ascidian Botrylloides magnicoecum (Qld)

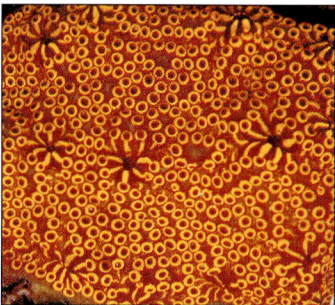
Leach's ascidian Botrylloides leachi (Mal.)

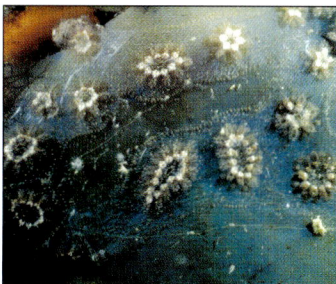
Investing ascidian growing by tunic expansion and budding (PNG)

Class: Chondrichthyes (Chond-ric-thy-es) sharks and rays

Modern sharks pre-date humans by 100,000,000 years and in this time, according to fossil records, they have not changed. Science estimates that the first ancestor of today's sharks appeared some 450,000,000 years ago in the Palaeozoic era.

There are around **340 species** of **sharks** in the **world's seas**, yet although most people can recognise their basic shape they only hear of tiger sharks or great whites as a result of shark attack and subsequent human fatalities.

All **sharks and rays** belong to a group known as **elasmobranchs** or strap-gilled fishes and they have a number of common features. The feature which separates them from other fishes is that they have **skeletons comprised of gristle**, either wholly **cartilaginous** or partly calcified.

Elasmobranchs do not have **scales**, and instead have an outer covering of tough skin which encases the entire body. This **skin** (once widely used as sandpaper) has countless, **tiny denticles** embedded in it. These **denticles** are actually **minute teeth** and injuries to humans have been caused merely by being brushed or bumped by a shark. The teeth of a shark are not embedded in the jaw but attached to the gums. When the outer teeth become worn or damaged, they are replaced with new ones. As most sharks swim, water passes through their mouth. **Oxygen** is extracted as the water passes over the **gills** and out through the five to seven gill slits. Some **species** can **rest** on the **bottom** and, by facing into the current, or by muscular control, pump enough water across the **gills** to provide for sufficient **oxygen** absorption

In general, most **elasmobranchs** have **spiracles**. These are body openings on each side of the head, either behind or to the side of the eyes. These openings lead to the mouth and are usually small in the free-swimming species and larger in the bottom-dwelling species, especially the rays.

Whale shark Rhinocodon typus (WA)

Galapagos shark Carcharhinus galapagensis (LHI)

Spotted wobbegong Orectolobus maculatus (NSW)

Elasmobranchs with flattened bodies (e.g. wobbegong sharks and stingrays) that **bury** in the sand or mud, actually **reverse** the **water flow**. Using their large spiracle openings on top of the head, these fishes draw in water from above which passes over the gills and is pumped out through the ventrally situated gill slits.

All sharks are carnivores, feeding on almost every kind of animal life. Although many species will feed during the day, most sharks hunt at night. Similar to other nocturnal hunters, they have special nocturnal eyes.

While the bigger sharks, the bottom dwellers and some whalers with distinctive markings are fairly easy to determine from sightings or a picture, many whaler species are very similar and require a specialist to distinguish them.

Manta ray Manta birostris (GBR)

White-spotted shovelnose ray
Rhynchobatus djiddensis (GBR)

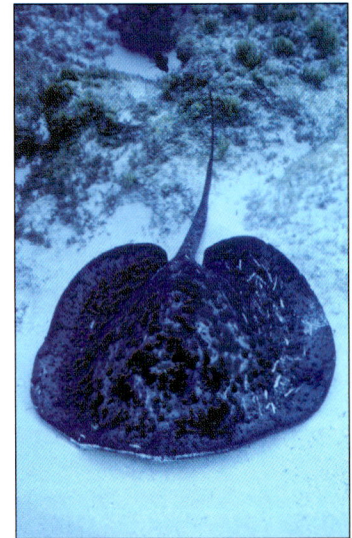
Meyen's stingray Taeniura meyeni (LHI)

Class: Osteichthys (Os-te-ich-thys) bony fishes

All the rayed-finned fish have bony skeletons and most have a single gill on each side of the head covered by a gill cover. There are some **5000 species** of marine fishes in Asia/Indo-Pacific waters.

Fish are the oldest known vertebrates and very distant relatives of humans. They are abundant in almost all aquatic habitats and have diversified to occupy almost all available **niches**. Together with developing to complement their habitat, fish have evolved to take advantage of the prey within each niche and many species have become specially adapted for this reason.

In each habitat or **micro-habitat**, the resident species generally conform to specific appearance. Once these become known it is possible to predict what type of fish will be found in certain habitats. This applies to either tropical or temperate waters. Although many body forms have evolved through natural selection and specialised feeding habits, in general, most marine fishes have four main body types:

1. **Laterally compressed:** bream, bullseyes, batfish, angelfish
2. **Dorsally compressed:** flathead, frogfish
3. **Rhomboidal:** tuna, tunny, bonito, kingfish
4. **Elongate, or snake-like:** moray eels, needlefish, trumpetfish

Bony fish feed on algae, almost every type of invertebrate and on other fishes. The law of the jungle applies equally as the law of the ocean - survival of the fittest, and the big fish eat the little fish.

Colour in fishes is keyed to the needs and requirements of the fishes lifestyle and behaviour, and just as the various life requirements change and alter so, too, can fish alter their colours and patterns to suit.

See **Indo-Pacific Sea Fishes (Northern Australia - Great Barrier Reef)** Neville Coleman for reference and more information.

Blue-ringed angelfish Pomacanthus annularis (Indo.)

Fringe-eyed flathead Cymbacephalus nematophthalmus (PNG)

Kingfish Seriola lalandi (LHI)

Pacific trumpetfish Aulostomus chinensis (GBR)

Leaf scorpionfish Taenionotus triacanthus (GBR)

Giant squirrelfish Sargocentron spiniferum (Mal.)

Green turtle Chelonia mydas (female) (GBR)

Green turtle Chelonia mydas (male) (GBR)

Green turtles mating (GBR)

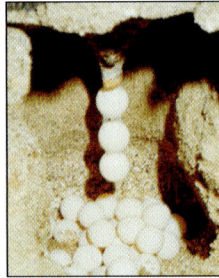

Female turtle laying eggs (WA)

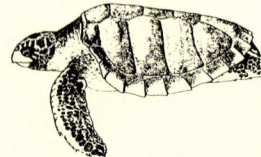

Loggerhead turtle hatchlings Caretta caretta (Qld)

Loggerhead turtle Caretta caretta (male) (GBR)

Class: Reptilia (Rep-till-ia) turtles and sea snakes

Order: Chelonia (Chel-on-ia) sea turtles

There are six species of sea turtles inhabiting the Asia/Indo-Pacific area. most are classed as endangered species.

Sea turtles have strong bony shells covered by large overlapping plates (the carapace). The limbs are well developed, heavy and paddle-shaped with one or more claws on the anterior edge and this basic design has not changed for over 150 million years. **Similar to all reptiles they breathe air** yet they are able to slow down their heartbeats and sleep underwater for several hours.

Marine turtles spend their lives in the sea, the only exception is that they are born on land and the females make short journeys up the beach to lay their eggs. Females have short tails and males have long tails (which aid in mating).

They feed on a variety of sea life including algae, barnacles, ascidians, sea jellies, fish, seagrass and crustaceans. Although they have survived throughout millions of years many species are now threatened, not only from increasing human predation but also from jettisoned waste in the form of plastic bags. The plastic bags (often bait bags thrown away by fishermen) may give off a marine smell (fish or prawns) and appear similar to a sea jelly. If eaten, the plastic blocks up the intestines and the turtle starves to death. (Discarded plastic bags are one of the greatest threats to marine turtles and carnivorous marine mammals.)

GREEN TURTLE   LOGGERHEAD TURTLE

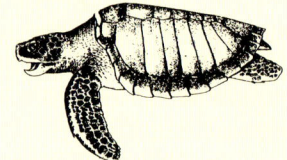

Order: Squamata (Squa-am-ata) sea snakes

Most sea snakes belong to the family Hydrophiidae and as **marine reptiles breathe air;** many have **valvular nostrils** and a vertically compressed, **paddle-like tail.** The body is covered in scales and most species are highly **venomous** and kill their prey by biting and injecting venom.

There are around 50 species of sea snakes known across the Asia/Indo-Pacific region and although most occur around the shores of continental islands and reefs, several species are known to drift and swim large distances across open ocean in the vicinity of sargassum rafts and driftwood.

Sea snakes **feed** on a variety of **reef fishes** and their **eggs**, and are able to hunt both day and night probing beneath rocks, in crevices, even under and around divers. They are especially keen on gobies and scorpion fish which are killed with venom and **swallowed head first.**

Although they are known to dive to depths in excess of 40 metres (131 ft) and sleep beneath coral slabs, or wrapped around coral for up to three hours they must return to the surface at various times to breathe. Sea snakes have poor eyesight and curious natures and for this reason they have been feared by reefwalkers, snorkellers and scuba divers. Like so many so-called "dangerous" sea creatures "prevention is better than cure". (See *Dangerous Sea Creatures - A Survival Guide* by Neville Coleman).

Olive sea snake Aipysurus laevis (PNG)

Olive sea snake. Closed nostrils (GBR)

Hunting sea snake with head down a goby hole (GBR)

Banded krait Laticauda colubrina is generally seen around reefs (PNG)

Diver with olive sea snake (WA)

Male claspers olive sea snake (GBR)

Saltwater crocodile Crocodylus porosus (NT)

Saltwater crocodile. Head and teeth (NT)

Spinner Dolphins Stenella longirostris (CS)

Bottlenose dolphin shark-bitten (Qld)

Bottlenose dolphin Tursiops truncatus (Qld)

Order: Crocodilia (Croc-o-dil-ia) crocodiles

Throughout its range the salt water crocodile *Crocodylus porosus* has been responsible for hundreds of attacks on humans. Though mostly nocturnal, this stealthy, cunning opportunist is bold and fearless when hunting prey, often venturing quite close to human habitation Most sightings occur in swamps and rivers, though some of the larger adults are regularly seen in open ocean waters and even on the Great Barrier Reef. After mating, females lay their eggs in a self-built humus mound or below the sand onshore and guard them until they hatch. Up to 60 young crocodiles may hatch from one nest. Crocodiles grow very large (up to eight metres/26 feet) have extremely thick skin, strong jaws and a powerful tail. In many parts of the world they have been hunted for their skins which can be processed into fine leather.

Despite their attacks on humans they are protected in some countries.

Class: Mammalia (Mam-al-ia) marine mammals

The word **mammal** refers to the female's **mammary glands** which provide milk for the young and are among the **main distinguishing features of a mammal.** Other features include hair, the birth of **fully formed live young,** ability to regulate body temperature and hold it constant, a brain of greater complexity than that of most other animals and breathing air. There are about 4070 species of mammals, many verging on extinction.

Ancestors of marine mammals are thought to have once lived on land. Returning to the sea, they have evolved to be the largest living creatures ever to exist on earth. Even so, by ignorance and over-exploitation humans have reduced many species almost to extinction in a mere 150 years. In some species of whales only a few hundred individuals remain from original stocks often in the vicinity of 100,000. Small isolated populations were decimated by hunting and many specialised forms are now extinct.

There are about 76 recognised species of **cetaceans** including all the **whales, dolphins** and **porpoises**. Of these, 66 have teeth (feeding on an array of prey ranging from fish to squid and, in the case of killer whales, other mammals), and 10 (great whales) have a system of hair 'plates' called baleen that is used to strain planktonic organisms (krill) and fish from the sea. All cetaceans are excellent swimmers, using both body flex and powerful tail flukes for their main propulsion.

Most of the commoner species of marine mammals can be readily identified from a good colour photograph of head, back and dorsal fin, or tail.

The Australian Government banned whaling in the same decade they officially recognised Aboriginals as human beings...

Whale watching; sustainable resource management (Qld)

Humpback whale Megaptera novaeangliae (Qld)

Southern right whale Eubalaena australis

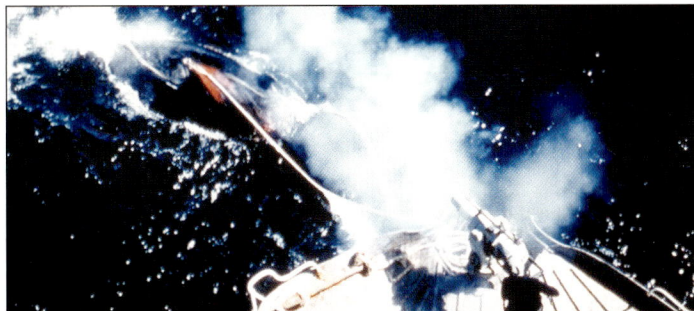

To really get the most out of diving and recognising the wildlife one needs to become a part of the environment rather than apart from it (GBR)

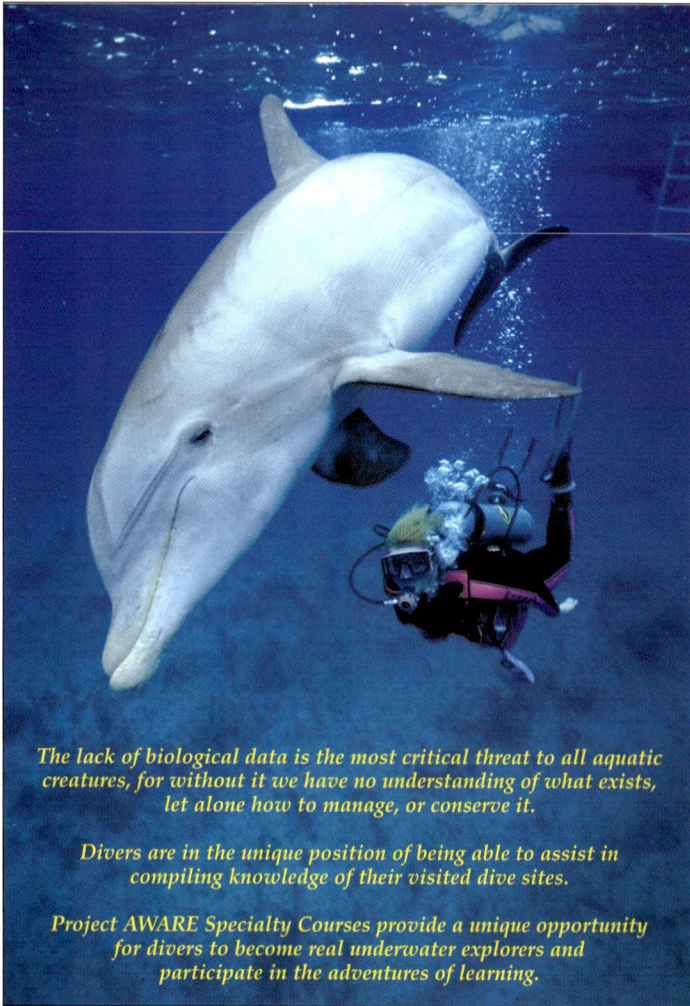

The lack of biological data is the most critical threat to all aquatic creatures, for without it we have no understanding of what exists, let alone how to manage, or conserve it.

Divers are in the unique position of being able to assist in compiling knowledge of their visited dive sites.

Project AWARE Specialty Courses provide a unique opportunity for divers to become real underwater explorers and participate in the adventures of learning.

PADI

What is Project AWARE?

Project AWARE (Aquatic World Awareness, Responsibility and Education) is a corporate environmental educational program of PADI worldwide.

As the world's largest recreational diving training and retailing organisation, PADI recognises its responsibility to preserve the aquatic environment. Project AWARE serves as a rallying point for PADI's almost 100,000 members and over 4000 dive centres/resorts worldwide. Through awareness, responsibility and education, PADI believes that divers can make a difference.

What is the Project AWARE foundation?

The Project AWARE foundation is the dive industry's lading non-profit (501(c)(3) organisation dedicated to the preservation of the aquatic environment.

We intend to teach the world about the importance and responsibility of preserving the aquatic environment by developing and disseminating educational materials, creating public awareness campaigns, promoting and organising industry efforts at a "grass roots" level, providing direct financial support for worthwhile endeavours, creating innovative projects and building alliances and partnerships with other organisations to strengthen our common goals.

Ten ways a diver can protect the underwater environment

Produced by PADI for Project AWARE

(1) Dive carefully in fragile aquatic ecosystems, such as coral reefs.

Although, at first, they may look like rocks or plants, many aquatic organisms are fragile creatures that can be damaged or harmed by the bump of a tank, knee or camera, a swipe of a fin or even the touch of a hand. It is also important to know that some aquatic organisms, such as corals, are extremely slow-growing. By breaking off even a small piece, you may be destroying decades of growth. By being careful, you can prevent devastating and long-lasting damage to magnificent dive sites.

(2) Be aware of your body and equipment placement when diving.

Much damage to the environment is done unknowingly. Keep your gauges and alternate air source secured so they don't drag over the reef or bottom. By controlling your buoyancy and taking care not to touch coral or other fragile organisms with your body, diving equipment or camera, you will have done your part in preventing injury to aquatic life.

(3) Keep your diving skills sharp with continuing education.

If you haven't dived in a while, your skills (particularly buoyancy control) may need sharpening. Before heading to the water, seek bottom time with a certified assistant or instructor in a pool or other environment that won't be damaged by a few bumps and scrapes. Better yet, take a diving continuing education course such as PADI Scuba Review, the PADI Adventures in Diving course or a PADI Specialty Diver course.

(4) Consider your impact on aquatic life through your interactions.

As every diver soon learns, very few forms of aquatic life pose a threat to us. In fact, some creatures even seem friendly and curious about our presence. As we become bolder and more curious ourselves, we may even feel compelled to touch, handle, feed and even hitch rides on certain aquatic life. However, our actions may cause stress to the animal, interrupt feeding and mating behaviour, introduce food items that are not healthy for the species or even provoke aggressive behaviour in normally non-aggressive species.

(5) Understand and respect underwater life.

Through adaptation to an aquatic environment, underwater life often differs greatly in appearance from life we are used to seeing on land. Many creatures only appear to look like plants or inanimate objects. Using them as "toys" or food for other animals can leave a trail of destruction that can disrupt a local ecosystem and rob other divers of the pleasure of observing or photographing these creatures. Consider enrolling in a PADI Underwater Naturalist or AWARE Tropical Fish Identification course.

(6) Resist the urge to collect souvenirs.

Dive sites that are heavily visited can be depleted of their resources in a short period of time. Collecting specimens, coral and shells in these areas can strip their fascination and beauty. If you want to return from your dives with trophies to show friends and family, you may want to consider underwater photography.

(7) If you hunt and/or gather game, obey all fish and game laws.

You may be among the group of divers who get pleasure from taking food from the aquatic realm. If you engage in this activity, it is vital that you obtain proper licensing and become familiar with all local fish and game rules. Local laws are designed to ensure the reproduction and survival of these animals. Only take creatures that you will consume. Never kill anything for the sake of killing. Respect the rights of other divers who are not hunting. Avoid spearfishing in areas that other divers are using for sight-seeing and underwater photography. As an underwater hunter, understand your effect on the environment.

(8) Report environmental disturbances or destruction of your dive sites.

As a diver, you are in a unique position to monitor the health of local waterways, lakes and coastal areas. If you observe an unusual depletion of aquatic life, a rash of injuries to aquatic animals, or notice strange substances or objects in the water, report them to local authorities, such as the local office of the Environmental Protection Agency or similar organisation in your country.

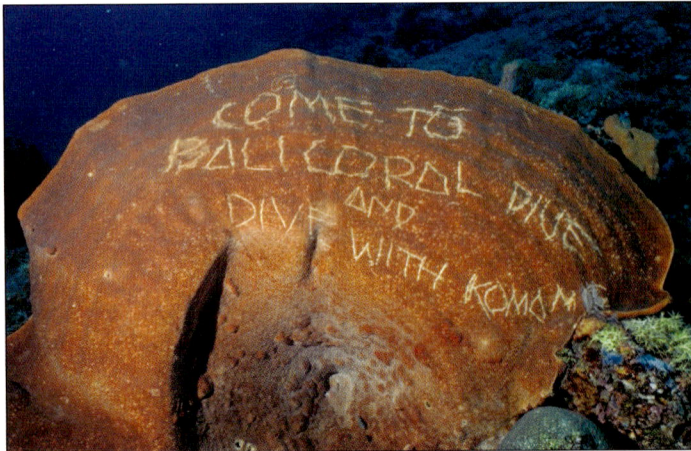

Visually polluting graffiti isn't good anywhere.
Underwater, carved into living organisms is doubly obnoxious.

(9) Be a role model for other divers in diving and non-diving interaction with the environment.

As a diver, you realise that when someone tosses a plastic wrapper or other debris overboard, it is not out of sight, out of mind. You see the results of such neglect. Set a good example in your own interactions with the environment, and other divers and non-divers will follow suit.

(10) Get involved in local environmental activities and issues.

You may feel you can't save the world, but you can have a great impact on the corner of the plant in which you live and dive. There are plenty of opportunities to show your support of a clean aquatic environment, including local beach clean-ups and attending public hearings on matters that impact local coastal areas and water resources. Know all sides of the aquatic environmental legislative issues and make your opinions known at the ballot box.

We need your support today...

Divers are conscientious protectors of the underwater environment. You can help preserve this precious resource for future generations by supporting Project AWARE (Aquatic World Awareness, Responsibility and Education). Order the limited edition Project AWARE version of your PADI certification card by adding a donation (minimum $5) to the Project AWARE Foundation when you replace your current card or complete your next PADI course   All contributions to the foundation are tax deductible, where applicable, and go toward beneficial programs such as environmental research, ecological clean-ups, and public environmental educational programs. Ask your PADI instructor, Dive Centre or Resort for further details.

www.projectaware.org

A View from the Board – Jean Michel Cousteau

Earth's natural resources are vital to the future of civilisation, and since humanity's quality of life is so closely connected to a healthy environment, we need to protect and properly manage our planet's resources. Over consumption, pollution and natural resource mismanagement undermine global habitability. I submit that it is our state of mind - our world view, beliefs and behaviour - that is at the root of these problems. The solution to our current environment crisis lies in preventing problems, not continually managing them. This requires an attitude adjustment on our part.

Too often we prioritise the individual first, then the community and finally the environment. We need to realise that we are only a small part of earth's living community and that we are inextricably connected to all life on earth - and vice versa. If we acknowledge our connection with the world, we can make environmental decisions today that will have positive effects on future generations. Together, we can discover new methods of sustainability for earth's natural resources.

Connecting with the environment captures our attention and enthusiasm and provides a positive setting for growth. The environment is a living laboratory that allows us to explore the structure, function and interdependence of life, as well as reproductive strategies, behaviours, ecosystems, diversity and evolutionary patterns.

We must expand our learning experience from basic biology and ecology to think about human kind's connection to our surroundings, and to our future. This surpasses scientific or utilitarian perspectives of the world in consideration of our connection to all life on earth. We need to focus on human interactions and how we relate to our surroundings. We are connected to every other species on earth by a constant flow of energy and life. Although we may look different, consider this: ever living organism on earth is made of the same building blocks from which all life flows.

Neville Coleman and Jean-Michel Cousteau have grown up as individuals, yet their basic philosophy evolved from a mutual understanding of universal principles (Photo: Barry Andrewartha)

Project AWARE
Honorary Board of Governors

ASIA PACIFIC	Neville Coleman
UNITED KINGDOM	David Bellamy
UNITED STATES	Charlie Beeker
	Patrick Cotter
	Jean-Michel Cousteau
	Steve Drogin
	Sylvia Earle, Ph.D
	Walcott Henry
	Sharon Lawrence
	Michael Newman
	Blu Rivard
	Roger T. Rufe, Jr.
	Dee Scarr
	Seba Sheavly
	Hillary Viders, Ph.D

Air pollution pollutes the entire world and then ends up in the sea

Erosion: The bones of the earth; headstone to humanity

Toxic waste: Dump it in the mangroves

Development: Transplant rainforest palms to beach; 'murder' by reclamation

Water pollution: Business as usual

CONSERVATION

As an underwater naturalist I believe that if we are to defend our **World of Water** and its resources against our own destructive influences we must concentrate efforts towards building knowledge into an effective weapon which can be utilised against our own ignorance and short-sightedness, such as this sample...

> "*As soon as a diver leaves the surface he begins to get cold and fuddled and stupid. If he is breathing air, his depth is limited because oxygen becomes toxic at high pressures and nitrogen has a narcotic effect. If he is to avoid the bends, his working time is limited by the need for slow recompression. His vision is commonly restricted to one metre or less and it almost never exceeds 40 metres. He cannot communicate without special equipment and all his equipment must withstand high pressures. The conduct of each dive is ruled by the physical and physiological limits which is always dangerous and sometimes fatal to succeed. Compared to a scientist at his bench, a diver is very inefficient indeed*".* (Woods & Lythgoe, 1971)

There is room and scope for study in many aspects of the **World of Water** but there must be a theme of cooperation between all parties of investigation and this can only be brought about by a revaluation of attitudes towards what the **World of Water** really is in relation to humans and what humans are in relation to this aquatic planet.

Perhaps even more than any other group of individuals on earth TODAY underwater people are linked closely to the sea and how it fares TODAY is our responsibility TODAY. We are the only ones who can see the changes, we are the only ones who can visualise long term effects. The majority of the public are unaware as to the need to protect the oceans, even more so than the land for, in most cases, even our abuse of the land eventually reaches the sea and doubles the cost. It doesn't really matter if we visit the **World of Water** for adventure, for treasure, for food, for science, or for fun, the very fact that we see, must make us aware of our responsibility towards the oceans.

As humans, we have only one enemy, ourselves. We can be lazy and self-righteous and blame the Government, the multi-nationals and every other conceivable bogey man; we can let the general apathetic attitudes our world is well known for, dictate terms, or we can take part in our own lives and eventually in reshaping a more concerned attitude towards our seas.

Let me quote from two books. One is very old and one not so old. One is common knowledge, the other can be heard every time we go down to the sea.

From the old (*the Bible*)

"Blessed are the meek, for they shall inherit the earth!"

From the not so old (*The Australian Beachcomber*, Neville Coleman 1978)

> *"The salty tang of the sea spray carried high by the updraught, zephyrs along on invisible winds and you can smell your air, your sea. It wafts into your ears and revives your brain. It soaks through your eyes and washes your soul. It infiltrates your very being, filling your air spaces, activating your forgotten senses. The goose bumps rise and your skin tingles with life; you can feel the pulsating rhythm as your blood pounds in time to vibrant thunder as a thousand tonnes of echoing turmoil heralds the arrival of another wave. The sea and its waves are older than life itself. Their rhythms are our rhythms."*

The meek, may inherit the earth.
But I think our seas...
are worth FIGHTING FOR.

Suburban and agriculture run off threatens the Great Barrier Reef

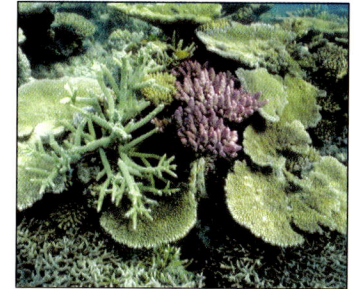

A healthy coral reef (GBR)

Globally warmed coral reef (GBR)

Education begins at the beginning

The sea and its waves...their rhythms are our rhythms

PHYSICAL LAND FACTS

(a) **Air** contains around 21% oxygen.

(b) **Oxygen** is a by-product of plant **photosynthesis** and is released into the atmosphere as an invisible gas.

(c) **Oxygen** is the life-support system for all life. (For most humans, **air** is taken into the body by instinctive breathing).

(d) **Air** is 800 times less dense than water.

(e) **Air** does not contain any visible life forms in its atmosphere.

(f) Most animal distribution is restricted to continents and islands. Birds, bats and insects are the only land animals capable of distribution by flying through **air**, across water. (Humans go by plane or boat).

(g) Wind currents transport and distribute reproductive spores, seeds and pollen of many thousands of plants through the **air**.

(h) **Air** is the medium by which the entire plant and animal kingdoms communicate. It carries the invisible scents and pheromones of chemical "conversation" so vital to the lives of all animals and plants. (To a lesser degree in humans).

(i) Plants can grow in suitable circumstances wherever **sunlight** penetrates. Plants cannot survive in the darkness, though fungi and some animals can.

(j) **Air** does not in itself cause differences in the spectral colours of **light**. (Human eyes may perceive the spectral values of **light** changing due to the various angles of the sun's rays at different times of the day).

(k) Most humans are unaware of **gravity** as a directional force in their lives.

(l) In order to grow, move, or stand erect within a gravitational force field, organisms must produce reinforcing structures such as wood and fibre in plants, or skeletons of bone and chitin in animals.

(m) High-energy lifestyles require a high percentage of carbohydrates (sugars/starch) in the diets in land organisms. (Humans can certainly relate to this).

PHYSICAL WATER FACTS

(a) There is around 6.5cc of oxygen in a litre of **water**

(b) Around 60% of **oxygen** in the earth's atmosphere is produced by aquatic plants not by rainforests as commonly believed.

(c) Oxygen as a gas is carried to the surface of **water** in bubbles, which burst on contact with the surface.

(d) With the aquatic environment, the percentage of **oxygen** distribution is variable depending on habitat zonation, e.g. high energy shores, where waves break continually, are high in **oxygen**, while the deep oceanic trenches and mangrove mud are low in **oxygen**.

(e) **Water** is 800 times more dense than air.

(f) Aquatic animals such as fish, molluscs, worms, crustaceans etc. have special types of gills which allow them to extract dissolved **oxygen** from the **water**.

(g) Aquatic reptiles (e.g. turtles) and mammals (e.g. dolphins) must return to the surface in order to breathe.

(h) For humans to survive underwater they must act in the same fashion as an aquatic mammal (snorkelling) or have some type of self-contained, or surface-provided **air** sources - scuba, hookah, or oxygen rebreather.

(i) **Air** supply to underwater humans is by way of a regulator, which only provides air on demand, requiring physical awareness and mental consciousness.

(j) Because **water** itself has no physical barriers as such, the distribution of animals and plants by planktonic dispersal of eggs and larvae by the currents is wide-ranging.

(k) **Water** supports an enormous variety of planktonic organisms living within its vertical column (sea surface to sea floor).

(l) Many aquatic animals are sessile, having no need to hunt as they are continually bathed in "living soup" brought by the currents e.g. feather duster worms, gorgonians, hydroids, corals, sponges, sea squirts.

(m) Currents also carry and disperse the "smells" and chemistry of the **water** world, indispensable to the survival of most aquatic creatures as the greater number do not have eyes and rely on "sniffing" the currents to find food and sexual partners.

(n) Humans have no senses other than sight, touch and sound by which to access **underwater**.

(o) Even in clear **water, sunshine** only penetrates to depths of around 183 metres (600 ft). Aquatic plant life is restricted to this area.

(p) With the absence of **sunlight**, some animals can either produce their own, or have **symbiotic relationships** with **light-producing bacteria**.

(q) In the absence of light, bacteria living in mangrove mud and in the hydro-thermal vents of abyssal trenches produce **chemosynthetic** energy which in turn supports other communities of animals.

(r) **Water** absorbs sunlight, especially the red end of the spectrum, which fades away in relatively shallow water, while blues and greens penetrate furthest.

(s) The vertical zonation of aquatic plants is determined by which part of the light spectrum is being utilised for **photosynthesis**.

(t) Due to the density of **water**, gravity does not exert any noticeable force, as it is offset by the buoyancy gained by an object's displacement of water.

(u) Apart from giant kelps, marine plants are generally low profile and pliant. None have the upright, structural tree-like growths of trunk and branches we are familiar with on land.

(v) In **water**, tree-like structures are built by **colonies of animals** e.g. corals, hydrocorals, black corals, gorgonians & hydroids.

(w) Slow, vertical movement in water requires little energy to be expended (humans also can relate to this).

(x) Most aquatic mammals in general have large stores of fat due to the high energy levels of their lifestyles e.g. diving to great depths for food, long migration patterns, keeping warm etc.

(y) Cold-blooded aquatic animals on the other hand generally require less carbohydrates as they do not store large amounts of fat. In contrast they are very high in protein. The high consumption of sea food by humans exploits this highly beneficial food source.

LOCALITY ABBREVIATIONS

(Fiji)	Fiji Islands	(WA)	Western Australia
(Indo.)	Indonesia	(Qld)	Queensland
(Jap.)	Japan	(NT)	Northern Territory
(Mal.)	Maldives	(GBR)	Great Barrier Reef
(PNG)	Papua New Guinea	(VIC)	Victoria
(Phil.)	Philippines	(SA)	South Australia
(Sey.)	Seychelles	(Tas.)	Tasmania
(Sol. Is.)	Solomon Islands	(Nor. Is.)	Norfolk Island
(Van.)	Vanuatu	(LHI)	Lord Howe Island

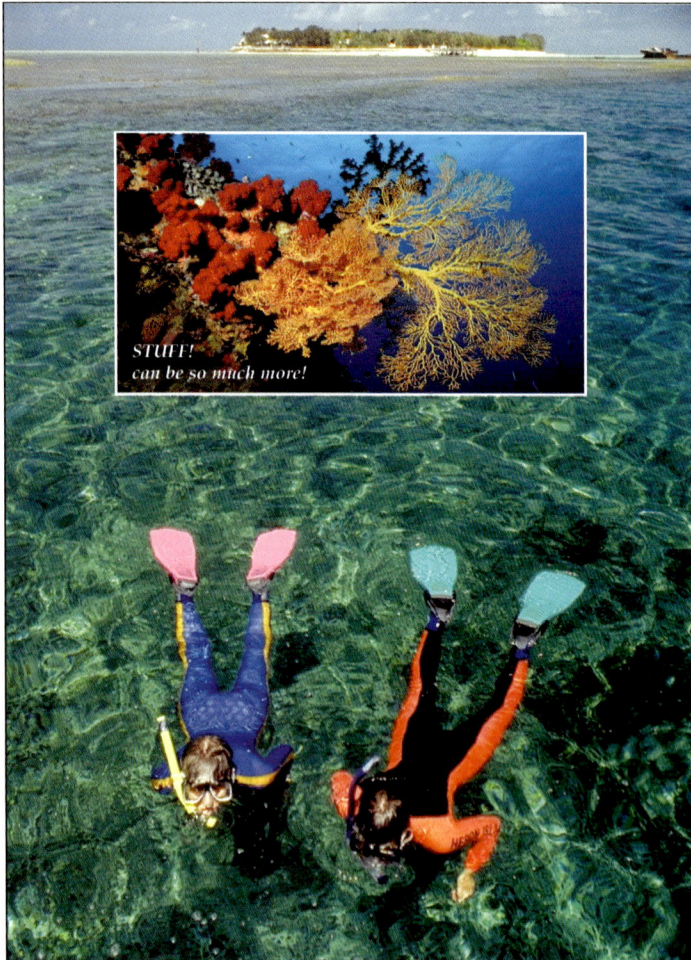

Eco-tourism and educational programs are some of the best ways for people to gain an appreciation of nature (GBR)

GLOSSARY OF TERMS

Asexual: reproduction by means other than sexual action, for example, by budding or splitting from the parent body

Autotomy: the spontaneous casting off of part of an animal's body, often to facilitate escape; for example, breaking off a captured limb

Branchial: respiratory function of an organ (gill) or region of body

Buccal: relating to the oral cavity (mouth)

Budding: the process of polyp duplication in the forming of a colony

Byssus: a tuft of strong filaments or thread-like strands with which some bivalves attach themselves to objects, such as rocks

Calcareous: composed of or containing calcium carbonate

Calice: the opening of the corallite, bounded by the wall

Carapace: chitinous and/or calcareous skin fold enclosing part or whole of the dorsal part of crustaceans and turtles

Carnivorous: flesh-eating

Cerata: tentacular processes on the backs of some nudibranch molluscs

Chela: the prehensile nipper or claw of some arthropods such as crabs; plural chelae

Chemoreception: sensitivity to water or airborne chemicals, especially developed in marine animals

Chitin: a horny organic compound forming part of the skin or shell of some marine animals

Ciliary motion: movement caused by the microscopic threads borne on the outer membranes of cells (cilia) of some marine animals which beat back and forth

Cirrus: a slender appendage; plural cirri

Cloaca: common opening for respiratory, reproductive and anal systems in a number of animals

Colonial: pertaining to communal animals of the same species living together, sometimes with organic attachments to each other, as in corals, or sometimes social links, as bees or wasps

Colonial organisms: organisms that live together in social or structural colonies

Commensal: a term applied to two species living in close association with one another, neither one at the expense of the other

Conspecific: of the same species

Corallite: skeleton of an individual coral polyp

Detritus: accumulation of dead animal and plant tissue and fine sediment, usually found on the sea floor

Dioecious: having separate sexes

Dorsum: the back or top (dorsal) surface

Endemic: native to and restricted to a particular locality

Endoecism: an habitual relationship between two animals where one takes shelter in the tube or burrow of another

Eversible: capable of being turned inside out

Fission: reproduction by splitting of a body into two or more parts

Flange: a lip

Foot: muscular extension of a mollusc's body used for locomotion

Gastropod: a class of molluscs including snails, having a shell of a singular valve and a muscular foot

Genus: rank in taxonomic hierarchy: group of animals or plants with common characteristics and origins, usually containing more than one species

Gregarious: found together in groups

Herbivorous: plant-eating

Hermaphroditic: having both male and female reproductive organs in one animal

Intertidal: between the extremes of high and low tides

Invertebrate: animal without a backbone

Mantle: an outgrowth of the body well which lines the shell in molluscs

Nematocyst: a coiled thread that can be projected as a sting from the cnidoblast cells that contain it (cnidarians)

Notochord: rudimentary spinal cord found in protochordates (ascidians)

Obligate: dependent

Operculum: lid or stopper; for example, a plate on the foot of a gastropod that closes the aperture of the shell when the animal is retracted

Papilla: a small projection extending from the body tissue; plural papillae

Pelagic: inhabiting open waters of oceans or lakes or the water column by swimming or drifting: not living on the sea bottom

Phylum: a primary taxonomic division of animals and plants

Pinnate: having branches on either side of an axis

Plankton: animals or plants, especially minute or microscopic forms, that drift suspended in seas, rivers, lakes and ponds

Polyp: a sac-like individual within a group of animals such as coral It can be of solitary or colonial existence

Proboscis: prehensile snout of a mollusc

Radiole: respiratory and/or feeding tentacles of some tube worms

Radula: a ribbon-like tongue bearing rows of teeth with which a mollusc reduces food to digestible particles or drills through other shells

Rhinophore: sensory tentacle on the head or anterior section of the mantle of opisthobranchs (Phylum Mollusca)

Rhizoids: slender hairlike structures that function as roots in mosses, ferns, fungi and related plants

Sedentary: immobile; refers to animals that remain attached to a substrate or that are unattached but do not move

Sessile: attached by the base (generally to a substrate)

Seta: bristle; plural setae

Siliceous: composed of or resembling silica, a glass-like material

Species: a group of individuals closely related in structure, capable of breeding within the group, but not normally outside it

Spicule: a minute, hard, needle-like body found in some invertebrates, such as sponges, soft corals, sea fans, and sea cucumbers

Subspecies: a geographical or other subdivision of a species that is sufficiently different to be recognised as such; a race

Substrate: the sea bed on which animals and plants live or are attached; rock, coral, mud or sand

Subtidal: below low tide level

Swimmeret: an abdominal limb or appendage adapted for swimming (of a crustacean)

Synonymy: collective names that designate the same species

Test: the hard covering of some invertebrates such as crustaceans, sea urchins

Type specimen: the original specimen from which a species was described

Veliger: the free-swimming larva of many molluscs

Vertebrate: an animal with a backbone

Viscera: the intestines

Water-column: area of water between the sea floor and the surface

Zooid: an individual forming part of a colony and produced asexually by fission; this term is often used in place of 'polyp'

There are still a myriad beautiful places on earth. We have a duty to protect them (PNG)

Underwater naturalists can play an important part in advancing knowledge (NSW)

INDEX

alcyonarian...43
anemonefish...47
Angiosperms...30
antennae..62
Anthropomorphism....................................26
armour...59
asexual...51
baleen..83
Bilateral symmetry......................................75
bioluminescent...46
Brachiopod...69
Bubble shell..66
budding...51
calcareous...36
carapace...60, 62, 63
carbohydrates...28
carbon dioxide..28
carnivorous...54
cartilaginous...77
cetaceans...83
chelae..64
chlorophyll..28
Cirrus...71
cloning...51
colonial ascidians.......................................54
colonies...49
Conservation...88
cuttlebone...68, 69
Cuvierian organs..75
Cyanobacteria..27
denticles..77
dichromatism..61
Dolphins..83
Echinoderms...71
elasmobranchs..77
exhalant..35
exoskeleton...59

fluorescent...51
Gastropoda..65
Hermaphrodites....................................54, 59
inhalant...35
insect...59
intertidal...72
invertebrates...53
Jean-Michel Cousteau................................87
lace corals..70
matrix..76
Medusae..37
Metamorphose......................................56, 59
micro-atolls..50
microalgae..32
Nematocyst...
nocturnal...55, 6, 62
nudibranchs..66, 67
nutrients..33
octocoral...43, 44, 46
omnivorous...72
Opisthobranch..66
oscula..35
Pentamorus...71
pharynx...76
phosphates..33
photosynthetic.............................28 - 30 - 33
phytoplankton..32
pinnate..43
plankton..48, 59
Pleurobranch..66
pneumatophores...31
polyp..37, 43
polyzoans..70
predatory...54
proboscis...58, 66
Project AWARE...
Prosobranchia...65

proteinase...52
protozoans..34
pycnogonid...58
radial symmetry..37
Radial symmetry...74
radula..66
raptorial claws..60
red tides...33
rostrum..61
scavenge..72
sea lilies...71
Sea mosses...70
sedentary...56
siliceous..36
skeleton...43
skeleton...74
spicules...36, 43
spicules..75
spines..74
spiracles..77
splitting...51
spongin...37
Stromatolites..27
swimmerets...59, 60
symbiotic...35, 43
temperate..50
tentacles..51
tests...76
toxic...33
tropical..50
tube feet..72
venomous..42
viviparous...32
whales...83
Zooid...70
Zooxanthellae.......................................43, 47

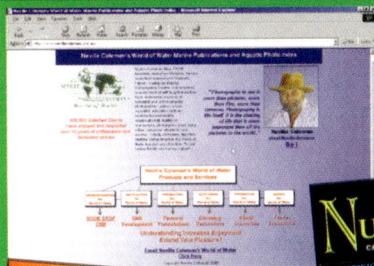

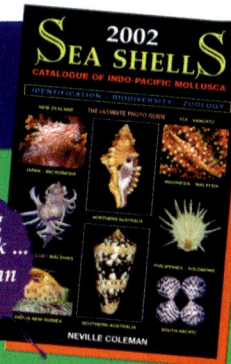

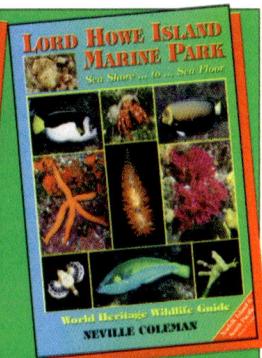

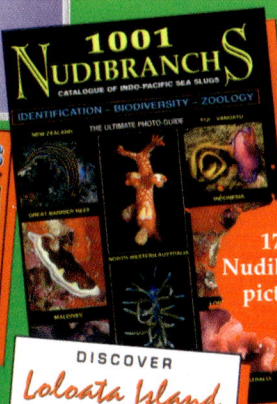

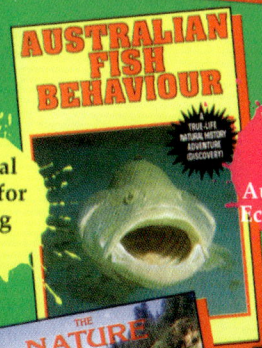

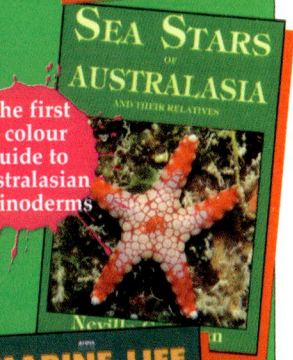

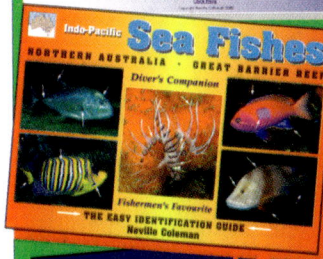

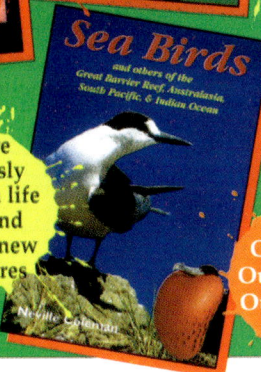

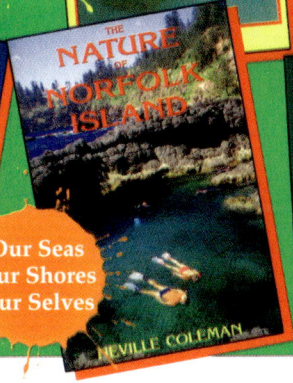

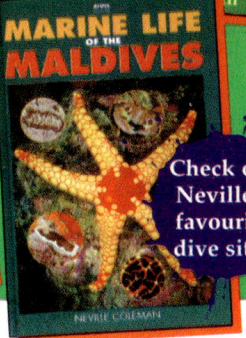

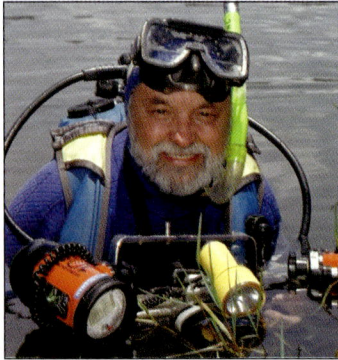

Neville Coleman Hon. FAIPP, Associate Australian Museum, Honorary Consultant Queensland Museum, is one of the most accomplished underwater natural history authors in the world.

Neville Coleman was born near the shores of the Lane Cove River in Sydney. As he grew up, fishing became his all-consuming passion. At ten years of age his most ardent aspirations were to become an explorer.

On leaving school he completed an apprenticeship in photo-lithography, but in 1963 his life reached a major turning point when, drawn by a love of nature and an unquenchable thirst for knowledge he set about to beat his greatest fear - the ocean and its inhabitants - and began spending his spare time diving in Sydney's harbour. His urge to discover, and the unknown challenge of the sea, eventually led to exploration on a larger scale.

In 1969, after two years of preparation, he conducted the "Australian Coastal Marine Expedition", a total of almost four years travelling 64,000 kilometres around the Australian coast, observing, recording, photographing and collecting many thousands of marine creatures. Most people who knew of the undertaking felt there was little chance of his getting back alive, let alone achieving the expedition's projected aims.

And so in March 1969 he set off - unfinanced, unsupported, unknown, undermanned and unlikely to succeed. However, succeed he did. It was to be the first underwater photographic fauna survey of an entire continent ever attempted.

Since 1973, Neville has cross-referenced approximately 150,000 transparencies with specimens of marine animals and plants donated to Australian museums.

Following the "Australian Coastal Marine Expedition", over 160 expeditions have been carried out in waters across the globe. Logging over 12,000 dives - and discovering over 450 species new to science, Neville's photographs are on display at most major museums and aquariums in Australia and at overseas institutions.

The Australasian Marine Photographic Index of which he is curator, is the largest scientifically-curated visual identification system in the Southern Hemisphere, with over 12,000 species photographed and catalogued.

In July 1980, London ATV flew Neville to Papua New Guinea for a 30-minute documentary in their Nature Watch series. This was eventually shown throughout Europe, USA and New Zealand with excellent reviews and proved to be one of the most popular in the series.

ABC's Big Country program also discovered Neville in 1980 and produced a 30-minute documentary on Neville's work at Lord Howe Island, Akin to the Sea, that also proved very popular and has since been reshown on ABC television in The Best of Big Country.

Early in 1985 Mike Willesee's Trans Media Production for Channel Nine Television produced a 1-hour documentary entitled Sink or Swim. In this program Neville introduces young institutionalised Australians to the wonders of underwater. In an effort to instil a positive side of life with his own personal philosophy of turning people on to nature, Neville explained how understanding the natural values of life can be one of the most rewarding methods of rehabilitation for body, soul and spirit.

Author of some 54 books, Neville has written and illustrated more four-colour underwater education natural history books than any other single person in the western world, and as such, is the most successful writer on marine life in Australia's history. His articles have been carried by over 150 magazines with photographs being reproduced by the National Geographic Society, Time-Life and Reader's Digest.

Neville is a fascinating and colourful individual with tremendous passion for life. He has an infectious enthusiasm for his work and has developed - through his experiences and knowledge - a confident understanding of the 'dangers' involved.

As the first full time professional freelance underwater naturalist / photographer managing to exist in Australia, Neville and his work are part of the pioneering spirit this country was built on, and for 14 years he single-handedly produced his conservation based "Underwater Geographic Magazine".

Neville lectures regularly throughout the world on underwater marine biology and conservation, and is certified by Australian dive instruction agencies to teach marine biology and underwater photography certificate courses. His "Education through Entertainment" audio-visual programs have been enjoyed by many thousands of people at over 300 world-wide venues.

His Nature Watch television program shown on National Geographic's Explorer series in 1986 has since been repeated several times reaching over 40,000,000 Americans. His lecture programs in the USA are highly successful and include conferences such as SEAS, MACNA, "Our World Underwater" and the Shedd Aquarium.

In 1991 Neville was awarded a number of prestigious honours including the Banksia Environmental Foundation's Marine & Waterways Award for his Australasian Marine Photographic Index, and the Diving Industry and Travel Association of Australia's Scuba Excellence Award for his contribution to underwater education. He also received an Honorary Fellowship from the Australian Institute of Professional Photography.

He is the first professional underwater photographer in Australia's history to win the highest commendation from both the Australian Photographic Society and the Australian Institute of Professional Photography.

In 1994 Brownies' Coastwatch (Channel Seven, Brisbane) hosted Neville as Marine Environmental Presenter on their weekly programs.

Vitally concerned with the aquatic environment and its conservation, Neville continues his exploration and discovery giving regular presentations to groups such as the Royal Geographic Society Explorer's Club. His appointment some years previously to PADI's "Project AWARE" Board of Governors is especially significant in his role as an environmental educator.

By taking the dreams of a ten-year-old and making them come true, Neville has already achieved more than most. Out of a world of total fear, a little boy who didn't have a hero, built one. The boy built the man; together they explore the ocean's unknown and share its secrets with humanity.